THE BIOCHEMICAL SOCIETY
ITS HISTORY AND ACTIVITIES

© *1969 The Biochemical Society*

Made and printed in England
by Castle Cary Press Ltd., Castle Cary, Somerset

THE BIOCHEMICAL SOCIETY
ITS HISTORY AND ACTIVITIES
1911-1969

by

R. A. MORTON, F.R.S.

Published by
The Biochemical Society,
7 Warwick Court,
London W.C.1   *December 1969*

FOREWORD	7
FOUNDATION OF THE SOCIETY	11
The Foundation, 4 March 1911	11
Acquisition of the *Biochemical Journal*	16
Progress of the Biochemical Society	22
THE RISE OF BIOCHEMISTRY	25
THE LEGAL STATUS OF THE BIOCHEMICAL SOCIETY	33
AIMS AND ADMINISTRATION OF THE SOCIETY	37
Proceedings	43
Accommodation of the Society	43
FINANCIAL AFFAIRS	49
GROUPS	55
THE BIOCHEMICAL JOURNAL	59
OTHER PUBLICATIONS	69
Clinical Science	69
Essays in Biochemistry	71
Occasional Publication	72
MEETINGS	73
Ordinary Meetings	73
Discussion Meetings and Symposia	76
Symposia Titles	80
Symposia on Education and Developments	81
Colloquium on Biochemistry in Industry	86
RELATIONS WITH OTHER BODIES	91
Chemical Society Library	91
Biochemical Abstracts	92

RELATIONS WITH OTHER BODIES *continued*
   Biological Council - - - - - - - - 95
   Parliamentary and Scientific Committee - - - - 96
   British Association for the Advancement of Science - - 97
   Joint Meeting with the British Association - - - - 97
   The Founding of the Nutrition Society - - - - 99

INTERNATIONAL AFFAIRS - - - - - - 101
   First International Congress of Biochemistry - - - 101
   The International Union of Biochemistry - - - - 104
   European Biochemical Societies - - - - - - 109
   The Australian Biochemical Society - - - - - 112

NOMENCLATURE, NOTATION AND ABBREVIATIONS - 113
   Patents - - - - - - - - - - 118
   Information Exchange Groups - - - - - - 119

CELEBRATIONS - - - - - - - - - 121
   Fiftieth Anniversary Meeting - - - - - - 121
   Sir Charles Harington - - - - - - - 128

HONORARY MEMBERS - - - - - - - 131

MEDALS, LECTURES, FELLOWSHIPS AND CONFERENCES 133
   Fellowships - - - - - - - - - 134
   The Harden Conferences - - - - - - 135
   Sir Frederick Gowland Hopkins - - - - - 135
   David Keilin - - - - - - - - 138
   Sir Arthur Harden - - - - - - - - 140

ANNEXES - - - - - - - - - - 145

FOREWORD

The idea of writing the history of the Biochemical Society came originally from a founder member J. A. Gardner who for a period of 31 years served as the Honorary Treasurer of the Society. With the warm approval of his Committee he began to collect material but he died before completing the task. R. H. A. Plimmer, also a founder member, became responsible for the work and his History of the Society was published in 1949. Gardner and Plimmer were in a very real sense founders of the Society, and it is difficult to exaggerate the value of their services. An account of Gardner's career by W. G. Ellis appeared in the *Biochemical Journal* (1947, **41**, 321) and Plimmer's contributions were recorded by J. Lowndes (*Biochem. J.* 1956, **62**, 353).

Plimmer's history provided a detailed and vivid picture of a period which to new members must now seem very distant. Excerpts from his account are reprinted here (pages 11 to 23).

The object of a revised and later version must be to tell the story of the rise and progress of the Society and to explain how its organization, administration and activities have evolved. The outstanding changes since 1949 have all, in one way or another, widened the rôle of the Society and extended its commitments. It is in the nature of scientific societies that committee members, officers and editors come and go, but by and large what they do is not undone by their successors, although it is often developed and improved. In the case of the Biochemical Society those who have been most prominent in revising its constitution and enlarging its scope seem well satisfied with what has been done. Few of them however deny that opposition to change, in Committee or Editorial Board, was on occasion justified as well as successful! How much of the necessary debate and controversy should be preserved and in due course made known is not easy to decide. In some respect the Minutes are very discreet about dissension especially when it caused anxiety to those concerned.

In preparing this revised account of the History of the Society I have had the benefit of numerous conversations with former officers and members who were at one period or another especially active. I have also had many letters containing interesting recollections, particularly on matters that at particular times were highly controversial. The correspondence will be preserved and (unless the writers object) placed in the archives of the Society. It was with considerable reluctance that I decided against quoting freely from some of the letters but it is better perhaps to lose a little vigour than to stoke up fires that have nearly gone out.

The remark which recurred again and again in the recollections of the most senior members was to the effect that the Biochemical Society was for long an amateur society, run by amateurs for amateurs at their own expense, and very economically at that.

A. C. Chibnall, writing in the *Annual Review of Biochemistry*, 1966, **35**, Part I, recalls "The only outside commitment I had in those days (1929) was that of Committee Secretary to the Biochemical Society, H. Raistrick dealing in a similar capacity with the business connected with the public meetings. We ran together in harmony with the Treasurer, J. A. Gardner, for seven years, foregathering one afternoon each year to check the books and to dine later with Gardner as host. Although the Society was flourishing and its membership had passed the seven hundred mark, the *Journal* was eating up all our available cash, and as secretaries, our official attendances at meetings, even as far away as Aberdeen, had to be at our own expense. Raistrick and I between us knew almost every member except those few who lived abroad, and the Society to us was just a happy family with Harden and Harington shouldering all the burden of publication."

Editing, (as C. R. Harington has testified) was firm and thorough, but essentially amateur. There was a real desire to build up financial reserves 'for a rainy day' but hardly any members foresaw the growth potential of the Society or the broader administrative implications. Indeed few would then have liked the prospect had they been able to imagine it. The early post-war years saw the beginnings of change. Meetings became much larger, the *Biochemical Journal* grew and with it the Editorial Board. The membership became bigger and younger.

I recollect that when I came to serve on the Committee and later to become Chairman a change in aims was already discernible. The Society was not standing still, nor was the subject, and enthusiasts on the Committee were determined to 'get a move on'. The question which occupied other members of the Committee was not whether changes were necessary so much as whether the right changes were being proposed. I recall a discussion with F. C. Happold on a journey from Leeds to Wharfedale and back. We examined as best we could the new trends and decided that they were on balance opportune. I realized incidentally that Happold could drive impeccably, keep his pipe lit and talk at the same time!

We were of course by no means alone, indeed many members and a succession of officers were giving much thought to general policy. Some members can claim credit for valuable suggestions but generally decisions were taken only after proper scrutiny of the new ideas. This scrutiny was usually carried out in committee but many issues were referred to the membership at Annual General Meetings.

Although the Society has been democratic, a number of individuals holding office have made distinctive contributions. This has been true all along and in recent years the formation of the International Union of Biochemistry owed much to a handful of British enthusiasts led by Harington. Later the formation of the Federation of European Biochemical Societies owed much to this country with W. J. Whelan as a driving force. The Society has had many first rate and devoted Honorary Secretaries with some like P. N. Campbell successful as innovators.

The Biochemical Society has had several 'celebrations' the last one being its Jubilee in 1961. The next landmark will be the 500th Meeting and the present revision of the History is part of the celebration.

In preparing the history of the Society I have had much help from Miss Doris Herriott and Mr A. I. P. Henton. Mrs M. Hilditch sorted out and reduced to order a large body of information and prepared the typescript. The Chairman (Professor Haslewood) and the Secretaries kindly suggested improvements. I am most grateful to them all and to my correspondents all of whom went to considerable trouble to place their recollections in perspective.

J. A. Gardner, co-founder 1911
Honorary Treasurer 1913–1943

R. H. A. Plimmer, co-founder,
Honorary Secretary and
Treasurer 1911–1912,
Honorary Treasurer, 1913–1919,
Honorary Member, 1943

THE FOUNDATION OF THE SOCIETY

R. H. A. Plimmer's earlier account of the History of the Biochemical Society covered a period of 38 years from 1911 to 1949 and eventful as those years were very great changes have taken place in the twenty years that have since elapsed.

At the turn of the century F. Gotch left the Liverpool Chair of Physiology to become Waynflete Professor at Oxford and was succeeded by C. S. Sherrington. Liverpool had a separate course in Chemical Physiology for medical students as early as 1899–1900, the lectures being given by Dr A. S. F. Grünbaum. In March 1902 Senate approved a proposal to establish a Chair of Physiological Chemistry and regulations for the professorship were drawn up in April. In May 1902 Council decided that the occupant of the Chair "be not expected to undertake teaching for the existing medical degree course, but that he devote his time to teaching advanced students and original research". It was also decided that the Chair should be filled by invitation.

The committee which drew up a document on the duties and tenure of the Johnston Chair changed the title from Physiological Chemistry to Biochemistry and recommended the appointment of Dr Benjamin Moore of the Royal University of Ireland from October 1902. The Minutes—in copper plate hand-writing—recorded that Professor Moore's salary should be £375 per annum and half the fees of his students. As it had already been passed that he was not to teach medical undergraduates, the fees must have been exiguous.

The first decade of the century saw Sherrington and others already established at Liverpool; F. G. Donnan became the first British professor of Physical Chemistry and Moore the first professor of Biochemistry in the country. Part of the importance of this lies in the fact that Moore with the help of Edward Whitley founded the *Biochemical Journal* (see pp. 16 to 20).

Plimmer wrote out of personal knowledge and we cannot do better than reprint parts of his History of the Biochemical Society. Much of the following section is taken from his account.

"THE FOUNDATION, 4 MARCH 1911

"Several considerations led to the formation of the Biochemical Society. The various medical schools in London had increased, or were about to increase, the staffs of their physiological departments, with chemists to teach and organize sections devoted to physiological chemistry and in some schools pathological chemistry. There was one department of Biochemistry in England at Liverpool, a corresponding department at Cambridge, and a lecturer at Oxford. Similar posts were being instituted in Scotland, Ireland and the provincial universities. Two appointments in Biological Chemistry existed at the Lister Institute together with junior staff.

"The holders of these new posts were actively engaged in research, and usually presented their papers to the Physiological Society, whose meetings were often mainly devoted to physiological chemistry so that papers on physiology were crowded out. At International Congresses of Physiology papers on physiological chemistry were given in separate rooms. Only those papers dealing with pure chemistry were suitable for publication in the *Journal of the Chemical Society*. The new biochemists were always

F. P. Worley C. Lovatt Evans P. Hartley E. L. Kennaway H. W. Bywaters W. H. Hurtley
J. V. Eyre S. A. Mann C. Dorée J. K. Close H. J. Page S. G. Paine J. Golding W. Cramer
E.Mellanby W.Ramsden R.H.Plimmer J.A.Gardner F.G.Hopkins A.Harden C.J.Martin B.Dyer T.A

Group of Original Members taken at the 21st Anniversary Meeting,
University College London, 17 November 1933

welcomed amongst the physiologists. The chemists regarded them as physiologists.

"Numerous chairs of Biochemistry existed in the United States of America, also in Germany and elsewhere in Europe. Many professors of Physiology here and abroad devoted their research to physiological chemistry.

"Three journals of Physiological Chemistry were published in Germany: Hoppe-Seyler's *Zeitschrift für physiologische Chemie*, started 1877; Hofmeister's *Beiträge*, started 1901; *Biochemische Zeitschrift*, started 1906. In the U. S. A. the *Journal of Biological Chemistry* appeared in 1905. In Great Britain the *Biochemical Journal* was started in 1906.

"Besides the physiological chemists various other workers had interests in Biochemistry: botanists, agriculturists, brewers, public analysts, medical clinicians and pathologists.

"It was necessary to advance the status of Biochemistry, in all its connexions, in Great Britain; its development here should not lag behind its progress in other countries.

"With these primary considerations in mind, and after conversations with colleagues, J. A. Gardner and R. H. A. Plimmer decided to call together as representative a gathering of biochemists as possible. A postcard dated 16 January 1911 was sent out to over fifty friends and fellow-workers asking them to attend a meeting in the Institute of Physiology, University College London, on Saturday, 21 January 1911 at 2.30 p.m., to discuss the formation of a Biochemical Society. Thirty-two of those invited were present, fourteen sent replies indicating their adhesion and suggested others who would be equally interested.

"J. A. Gardner presided and gave the chief reasons for calling the meeting. He emphasized the growing importance of Biochemistry both on the animal and vegetable sides. The increasing number of workers rendered the formation of a Biochemical Society desirable to serve four main needs: (1) a common meeting place to discuss biochemical problems; (2) the association of the workers on the animal and vegetable sides; (3) a common journal to be owned by the society; (4) the advancement of Biochemistry in this country.

"W. D. Halliburton, in opening the discussion, was strongly in favour of the formation of such a combined society with its meetings on unconventional lines. He moved a resolution to this effect, which was seconded by F. G. Hopkins, A. E. Garrod and A. Harden.

"H. E. Armstrong, who was opposed to any specialization, said that the main object should be to have a 'focus point', and that a society or club wherein the social side of the gathering preponderated should be a primary condition. E. J. Russell, speaking for agriculture, said the number of scientific papers was not large, and thought they would be of more value if brought before other biochemists. E. F. Armstrong hoped no omission would be made of workers on the botanical side. Plimmer, in summarizing the subjects so far mentioned, said the chemistry of brewing came into consideration as well.

"Finally, it was proposed by Armstrong, seconded by Halliburton and carried unanimously, 'that provisionally a club be established to promote intercourse among those biologists and chemists who are mutually interested and concerned in the investigation of problems common to biologists and chemists'.

"To make preliminary arrangements, Halliburton proposed that there should be a small committee limited to the conveners of the meeting. As these two gentlemen did not sufficiently represent all the interests, a committee of five was chosen. It consisted of J. A. Gardner, A. E. Garrod, W. D. Halliburton, R. H. A. Plimmer and E. J. Russell. Plimmer was asked to be Secretary."

The Committee met on 18 January and 2 February 1911. It was decided to follow the practice of the Physiological Society, which met at various laboratories and had no appointed president, rather than adopt the formalities of the Chemical Society. With this simplicity of organization and with the adoption of rules very similar to those of the Physiological Society it was a comparatively easy matter to make a start. The first meeting was held on 4 March 1911 at University College London with W. M. Bayliss in the Chair. Nine papers were read and an adjournment was made for dinner.

The report of the Committee suggested that the number of original members should be limited to about seventy. H. E. Armstrong spoke strongly against the report, saying that matters had been hurried and that a society formed on the lines suggested would not give a substantial organization capable of promoting the interest of the subject and those interested in it. The present proposals had gone much too far. Biochemistry was the coming subject and would play an important part in the future. No limitation of the number of original members should be made, and until the association possessed a journal of its own it should remain a Club. After prolonged and argumentative discussion about the title Club or Society he moved that this association provisionally be called 'The Biochemical Club'. This was carried. The eligibility of women was next considered. Again after long discussion the meeting decided that men only be eligible as members of the Club. The exclusion of women fortunately did not last long; on 5 February Miss Harriette Chick (later D.B.E. 1949), Miss Ida Smedley (Mrs Smedley Maclean) and Miss M. Wheldale were elected. Dr Chick was the first woman to serve on the Committee (1918).

FOUNDATION OF THE SOCIETY 15

"The following committee for the session 1911–12 was elected; H. E. Armstrong, W. M. Bayliss, A. J. Brown, H. H. Dale, J. A. Gardner, A. E. Garrod, W. D. Halliburton, A. Harden, F. G. Hopkins, F. Keeble, B. Moore, W. Ramsden, E. J. Russell, R. H. A. Plimmer (as Honorary Treasurer and Secretary). The chosen members represented the various biochemical interests."

During the 1911–12 session meetings were held at Oxford, Rothamsted, City and Guilds College, Cambridge (School of Agriculture), King's College, London, Lister Institute and St. Bartholomew's Hospital. The first Annual General Meeting was held at University College London on 2 March 1912. It was reported that attendances at meetings had averaged 40, "considerably less at the dinner".

There was no rule about the retiring members of the committee. The Chairman put forward the proposal of the committee that two members retire on account of least attendance and one member on account of seniority. A. J. Brown and B. Moore (his absence was due to his resignation during negotiations about the *Biochemical Journal*) retired for the first reason and A. E. Garrod for the second (by ballot). An amendment to the effect that retiring members be selected by the Club was carried.

At this stage H. E. Armstrong resigned*. Though it was explained that this vote was not meant as a reflection on the committee, Professor Armstrong still refused to reconsider and rejoin the Club and committee and left the meeting. The motion was subsequently rescinded. A. J. Brown, A. E. Garrod and W. D. Halliburton were voted as the retiring members. J. S. Ford, W. H. Hurtley and J. Lorrain Smith, as proposed by the committee, were elected as the new members.

Finally, it was resolved that Armstrong be specially asked to reconsider his resignation. Plimmer went next day to see him and later E. H. Starling tried, but to no avail. "The Club can never be unmindful of the help, encouragement and advice that Professor Armstrong gave in its early days."

In the 1912–13 session meetings were held at University College, Reading (Botany Department), Woburn Experimental Fruit Farm, South Kensington (University Department of Physiology), University College London, John Innes Horticultural Station, Birmingham University (Brewery Department), Cancer Hospital, London.

The seventeenth meeting was the second Annual General Meeting and was held in the Chemical Department at University College London with

---

*Professor Armstrong was able and influential and could be cantankerous. The present writer heard him, as an old man, fulminating about what he felt was the dreary lack of style in the *Journal of the Chemical Society*. Armstrong ferociously enjoyed being (partly) right on many issues.

Hugh Candy in the Chair. The Secretary-Treasurer reported a balance of £2. 16s. 2d. after paying all expenses, including the purchase of the *Biochemical Journal*. The first number of the *Biochemical Journal* edited by Bayliss and Harden, appeared in January. J. A. Gardner was appointed Honorary Treasurer and R. H. A. Plimmer Honorary Secretary. The new members of the committee were elected.

"The number of members at 31 December 1912 was 148, after three resignations and the loss of one member by death. During this period of two years the lengthy negotiations about a *Journal* were dealt with by the committee.

"THE ACQUISITION OF THE BIOCHEMICAL JOURNAL

"At the outset it was realized that if the Biochemical Club were to become permanent and improve its status from an informal discussion and dining club to a real society it was essential for it to possess its own *Journal*. The stumbling block was not financial, but the existence of the *Biochemical Journal* edited by B. Moore and E. Whitley at the University of Liverpool.

"Professor Moore was a member and strong supporter of the Biochemical Club. The committee met Moore in consultation on 11 February

B. Moore, editor with E. Whitley of the *Biochemical Journal* 1906–1912

Sir William Bayliss, F.R.S.,
Editor of the *Biochemical Journal*
1913–1924

Sir Arthur Harden, F.R.S.,
Editor of the *Biochemical Journal*
1913–1937, Honorary Member
1938

1911. It was decided not to issue printed proceedings for distribution at the meetings. Moore offered to accept papers of members of the Club and act in conjunction with the committee in regard to their publication and proposed to issue the new volume under the editorship of B. Moore and E. Whitley with the collaboration of the committee of the Biochemical Club. The subscription to the Club should not include the *Journal*, but members would be able to obtain it through the Club at a discount of 15% on the published price. These terms were reported at the meeting of the Club on 4 March 1911.

"The proposal was not acceptable to the Club Committee which wanted a *Journal* of its own. Moore was to be asked on what terms he would hand over the *Biochemical Journal* to the Club. He met J. A. Gardner and R. H. A. Plimmer on 4 July 1911 and explained that he started his *Journal* because of his desire that contributions should be published as submitted without criticism or editorial suggestions. His view was that authors of poor papers would take the blame and not the *Journal*. He was prepared to transfer his *Journal* on this basis of free and unrevised publication. The cost of publication was about £150 a volume, and there was a deficit of about £200 which might be settled satisfactorily. Gardner and Plimmer pointed out that a rival journal would compete with Moore's journal and had a good chance of success as most workers in Biochemistry had joined the Biochemical Club; yet it might not succeed. Moore wrote four days later (8 July) to say that the Club should start its own *Journal*, and in order to give the committee freedom of action he resigned his membership of the Club.

"The committee on 8 July 1911 discussed the pros and cons of publishing. Some journals had guarantors who had never been called upon. It was believed it would be possible to publish a journal without loss. So W. Ramsden was asked to make inquiries at the Oxford University Press, F. G. Hopkins at the Cambridge University Press, R. H. A. Plimmer at the London University Press and at some private publishers and printers. They reported to the committee on 14 October 1911. Comparison of the estimates showed the cost to be from £170 to £200 a volume. A private publisher offered to take the whole responsibility without guarantee and give half the profits to the Biochemical Club.

"A suggestion of H. E. Armstrong that he with Plimmer and others act as guarantors, so that a journal be speedily published, and hand over the *Journal* when published to the Club was not received favourably.

"F. Keeble then moved that the *Journal* be published by a University Press, and that detailed particulars be obtained from the Oxford and Cambridge University Presses. Ramsden and Hopkins were asked to continue their previous negotiations.

FOUNDATION OF THE SOCIETY

"Ramsden and the Oxford University Press felt that in the interests of Biochemistry in this country two journals should not exist, and Ramsden again tried to get Moore's co-operation. Professor Moore sent a draft memorandum of his terms: a sum of £260 payable in four yearly instalments of £65, Moore and Whitley to remain as editors until the money was paid. The *Biochemical Journal* had 170 subscribers of whom twenty-four were members of the Biochemical Club. The Club committee was told later that the price represented one-and-a-half years' purchase at £1. 1s. a subscriber. This high price could not be accepted by the committee.

"H. E. Armstrong, though he considered it desirable to buy the *Biochemical Journal*, said that no more than £100 should be offered. Later at F. G. Hopkin's suggestion, he proposed that Principal Miers of Manchester University be asked to assess the value of Moore's *Journal* to the Club. Moore and Whitley met the committee and agreed to the valuation, but neither side was to bind itself to accept. Principal Miers agreed to act if a short statement of the negotiations with Moore were submitted to him. His valuation of £150 was reported to the committee on 20 January 1912. Ramsden was not content with this verdict and asked Moore to send his own statement to Principal Miers. He made no alteration to his valuation. The disparity was great and not pleasing to Moore.

"The Secretary reported to the committee that he had met Professor Moore in December and asked him if he would agree to a valuation by Mr W. M. Meredith of Messrs Constable and Co. The answer was 'yes'. Mr Meredith had agreed to act only if he could ask any questions, and that his award was adhered to by both parties. Moore wrote that he could not agree to the conditions.

"Finally, at this meeting of the committee, to overcome this deadlock, Keeble proposed that Principal Miers's valuation of £150 for the purchase of all rights in the *Biochemical Journal* as specified in the memorandum below be communicated to Moore as a definite minimum proposal from the Biochemical Club—'Should he be unwilling to accept the proposal the offer is made to refer to Mr Meredith for final adjudication, both parties agreeing to accept Mr Meredith's valuation as final.'

"MEMORANDUM

"1. In consideration of the terms contained in subsequent paragraphs the vendors, Messrs B. Moore and E. Whitley and the University Press of Liverpool, agree to hand over the *Biochemical Journal* to the Purchasers, the Biochemical Club, as a going concern and free from all debts together with a list of subscribers thereto standing at present at 170, but all copies of back volumes and numbers already issued of the current volume shall remain the property of the vendors.

"2. The Biochemical Club agree to pay forthwith to the vendors the sum of £150 in purchase of the goodwill and subscription list mentioned in clause 1 and to take over and be financially responsible for the issue and management of the *Journal* as from a date to be agreed upon.

"3. The *Biochemical Journal* shall be wholly and solely held, edited and managed by the Biochemical Club.

"If this offer now made to Professor Moore be not accepted the Biochemical Club proceeds to establish its own *Journal* independently."

"On 3 February 1912 two letters from Professor Moore stated that he agreed to accept the Biochemical Club's proposal to buy the *Journal* for a minimum price of £150, but he wished for an interview with Mr Meredith to see if the figure could be raised. He desired (1) to have the first option of recontinuing the *Biochemical Journal* if for any reasons the Biochemical Club ceased to publish it, and (2) that the title should not be changed and the volumes renumbered as from the taking over. Professor Moore was informed that the Committee could not accept the limitations of the second point, and that he should give a statement that he agreed to the original terms.

"On 2 March 1912 the Secretary informed the Committee that Professor Moore was unable to meet Mr Meredith and had written agreeing to accept the valuation of £150.

"The arrangements were thus at last complete, and it was decided to take over at the completion of the current volume (number 6). Professor Moore would state in his next number that in future the *Biochemical Journal* would be issued by the Biochemical Club.

"Later, he inserted a slip repeating this information, setting out the objects of the Biochemical Club, and stating that the subscription was 25s. per annum including the *Journal* for the year. Other subscribers were asked to pay £1. 1s. per volume.

"THE PAYMENT OF £150

"Before the negotiations with Professor Moore were completed, a generous gift of £25 from Professor Sir William Osler (through Dr Ramsden) was gratefully accepted. There was a deposit of £40 and a balance of £6 from the first year. A similar balance of £40 was expected from the second year. The Secretary felt that members would like to give donations and feel that they had helped to buy the *Journal* for the Club. In this way £30 was subscribed. A gift of £5 from Mrs Herter was sent from New York through H. D. Dakin. Dr Vincent kindly contributed the last £10.

"An agreement for the purchase was drawn up by a solicitor. The Chairman of Committee (A. Harden) and the Secretary (R. H. A. Plimmer) were authorized to sign the deed of assignment, and the Secretary authorized to pay £150 to Professor Moore and Mr Whitley.

"The Biochemical Society and the *Biochemical Journal* are now so well and firmly established and taken for granted that few of the present members know anything of the troublesome negotiations which harassed the committee of the Biochemical Club during the first two years.

"FINAL ARRANGEMENTS

"Some additional details were still necessary. A sub-committee consisting of J. A. Gardner, A. Harden, F. G. Hopkins and the Secretary was appointed to report on (1) the title and constitution of the association, (2) the cost of publication of the *Biochemical Journal*, (3) the amount of subscription, based if necessary on a canvas of members.

"The sub-committee, in view of past argumentative discussions at annual meetings, decided to take a poll by postcard on three questions:

(1) Is it your opinion that membership of the Club should involve compulsory subscription to the *Biochemical Journal*?
Answer: Yes 65; No 25.

(2) In the event of the subscription to the *Biochemical Journal* not being compulsory for all members, are you prepared to subscribe to the *Journal* at a cost of 15s. to £1 per annum, in addition to the present subscription to the Club?
Answer: Yes 72; No 19.

(3) Are you in favour of changing the name of the association to 'The Biochemical Society'?
Answer: Yes 79; No 10.

"It was clear that the subscription to the *Biochemical Journal* should be compulsory for all members and that the title should be 'The Biochemical Society'.

"The tenders for printing the *Biochemical Journal* showed that the most favourable terms were those of the Cambridge University Press: £200 approximately for an issue of 500 copies of eight parts of 80 pages per volume in the style of the present *Journal*. It was estimated that a subscription of £1 per member would cover the cost of publication. Under the title of the *Journal* the words 'edited for the Biochemical Society' should be inserted.

"EDITORSHIP OF THE BIOCHEMICAL JOURNAL

"The Committee decided that the *Biochemical Journal* should be edited by two editors and a representative editorial committee.

"No definite record exists of how the first editors were chosen. The Secretary (Plimmer) well remembers how he thought that one editor might represent the more chemical side and the other the more physiological, and that if he could secure the services of Harden and Bayliss as editors the greatest benefits would come to the Biochemical Society and the *Biochemical Journal*. He made special visits to Doctors Harden and Bayliss and was agreeably surprised and overwhelmed with delight to learn that both would accept. It was the finest possible culmination to all the work in connection with the Biochemical Society.

"PROGRESS OF THE BIOCHEMICAL SOCIETY

"Since the foundation of the complete society with its own *Journal* in 1913 there has been continuous development which is best described under different headings.

"MEMBERSHIP

"Fewer members than expected of the original Biochemical Club resigned on its being transformed into a Society. In the first few years, even during the war years of 1914–18, there was an average yearly increase of eight or nine. A more rapid increase in numbers started in 1920 in spite of losses by death and resignation. For many years the increase averaged about fifty. In one year it was over eighty. The number exceeded 1000 (1017) on 1 January 1944. In one of the later years over 100 new members were elected. At the Annual General Meeting in March 1949 the membership was reported to be 1398. 'The founders and original members hardly contemplated such a large number, and those surviving cannot but be highly gratified with the great success achieved during these 36 years."

MEETINGS

Plimmer's account goes on:

"The original practice of holding meetings at various colleges and medical schools in London and at other centres outside London with the Head of the laboratory as Chairman has been continued throughout the 38 years of the existence of the Biochemical Society. The informal character of the meetings has been maintained.

"The actual procedure was altered at the annual meeting in 1913. The minutes, announcements, business and communications were taken first instead of after dinner, so that more members were present to discuss any special matter. The attendance at dinners was small. The social side was kept separate and a Biochemical Dining Club was formed. It was not well supported and was discontinued after 15 May 1915. Still the social side was not neglected. Arrangements where possible were made for members and guests to have lunch or dinner. During the last years meetings have been held on days preceding the meetings of the Physiological Society so that members of both societies could sometimes attend a joint meeting at dinner.

"Eight meetings a year were considered a suitable number. During the war years of 1914–18 the number was less and six to eight were arranged. Again, in the years 1939–45 meetings, though fewer, were held in spite of the wartime difficulties. The Annual General Meeting in March 1949 was the 175th meeting to be held.

"The attendance at the meetings from 1913 to about 1921 averaged from 40 to 50. It became greater with the increase in membership, and at times in the last years has reached over 200. The accommodation in some lecture theatres has been taxed to the utmost. At some meetings the communications were given in two sections, and an extra meeting was inserted in 1948–49. As the communications also largely increased a time limit was fixed. Two to three hours was sufficient time in the early years; half days and whole days have been necessary recently."

DISCUSSION MEETINGS

Discussion meetings were an early feature of the Society's work. Plimmer noted a session on micromethods of analysis (December 1914) another on the analysis of meat extracts (March 1915). The war delayed matters and the next discussion meeting was a discourse by R. Stenhouse Williams (June 1918) on the commercial production of uncontaminated milk. This was followed by a joint meeting with the Physiological Society on 'The validity of the isodynamic law of nutrition of interavailability of fat and carbohydrate' (November 1918). Another break occurred until (February 1921) an informal discussion was held on 'The internal structure of the protein molecule and its bearing on the chemical, physical and biological properties of proteins' (W. M. Bayliss in the Chair).

New ground was broken at Edinburgh when G. Barger was in the Chair at a discussion on the place of biochemistry in medical research and education (September 1921).

A joint meeting with the London Section of the Society of Chemical Industry was devoted to 'Micro-organisms and their application to industry and research in organic chemistry' with Chaston Chapman in the Chair (January 1923) and in May 1923 a joint meeting with the Institute of Brewing discussed 'Biochemical aspects of fermentation' with Harden in the Chair. Three years later another joint meeting with the Society of Chemical Industry tackled 'The scientific and industrial problems presented by the hormones'. The speakers included H. H. Dale, H. W. Dudley, F. H. Carr, H. A. D. Jowett and G. Barger with Sir Alfred Mond in the Chair (July 1926).

The joint meetings with the Society of Chemical Industry included (July 1927) a discussion on 'The physiological and industrial aspects of the chemistry of the carbohydrates'.

In October 1928 Hopkins gave a special lecture in commemoration of the centenary of Friedrich Wöhler's synthesis of urea. There was then a prolonged break until November 1941 when a discussion was held on the mode of action of chemotherapeutic agents with E. C. Dodds in the Chair.

Plimmer did not record the fact that revival of Discussion meetings was proposed before the War and that the initiative was taken by H. J. Channon in 1934 (see page 76).

THE RISE OF BIOCHEMISTRY

It is impossible to understand the history of the Biochemical Society without some appreciation of how biochemistry itself has grown during the past 60 to 70 years. Equally, the development of the subject cannot be isolated from the parallel advances that have been made in chemistry and physics as well as in all the biological sciences. It is often said that the 20th Century has been swept along by new scientific ideas and (some would think) swept off its feet a little by technology. The boundaries between disciplines and technologies have been blurred and 'territorial rights' have become obsolete.

Over the same period some of the applications of science, and not least of biochemistry, have profoundly influenced the general welfare. In recognition of the benefits already obtained and in the hope of further intellectual and material gains the financial provision for biochemical teaching and research has enormously increased. Students in many countries find the subject attractive and university departments of biochemistry have increased in number and size, particularly in the last 10 years. The output of published research has risen rapidly and scientific societies have multiplied. New journals have been founded and have paid their way, largely because of world-wide institutional subscriptions. It is in this setting of extremely rapid growth that we have to place the recent history of the Biochemical Society.

If we look back to the first decade of the century when the *Biochemical Journal* was founded by Benjamin Moore and Edward Whitley as a brave private venture, there is nothing parochial about the contents of the early volumes. Any contribution offering a prospect of linking biology and chemical molecules was welcomed. The topics ranged from glycosuria to trypanosomiasis and from colloid chemistry to enzymology and took in on the way renal calculi and some marine biology. Benjamin Moore and his group had a vision of Biochemistry in breadth just as Sherrington, working next door, had his insight into Physiology in depth.

If today we contemplate *Annual Reviews of Biochemistry*, the thirty volumes of *Comprehensive Biochemistry* and the seventeen volumes of *Methods of Biochemical Analysis* now on our bookshelves, it cannot be denied that biochemists accept few restrictions on subject matter and that their capacity to assimilate techniques—irrespective of provenance—is prodigious.

In advance of their time in ideas, the founders of the *Journal* were on the crest of a wave—apart from the shortage of money, although the

Liverpool building and the Chair had been privately endowed. In a preface dated December 1911 to Volume 2 of the collected papers, Moore wrote of his department:

> "There are rarely less than ten research workers in the laboratory at a given time and after the unavoidable expenses of cleaning, gas, light and water have been met, there remains to defray the cost of materials for research an annual sum of about one hundred and twenty pounds. Even when the most economical types of work were chosen the average annual expense of a research worker in a laboratory works out at a sum of at least twenty pounds. It follows that for about one third of each year all funds are lacking for the purchase of the most ordinary laboratory materials. Such austere and rigid economy is altogether incompatible with first-class creative work".

G. A. D. Haslewood, Present Chairman of Committee

Although there are still parts of the world where this cry can be echoed, in most places continuing inspiration is much better served in a material sense.

In Britain the turning point came in 1913 when the Government established the Medical Research Committee, with the duty of advising how best to spend on medical research £40,000 to £50,000 per year, a sum accumulated from the annual contribution of one penny by each person insured under the National Health Insurance Act of 1911. [A rough correction for inflation puts this as equivalent to at least £500,000 today.] As E. Mellanby wrote (*A Story of Nutritional Research*, (1950) Baltimore, Williams and Wilkins):

> "This was the beginning of the State's financial interest in the serious promotion of medical research, and indeed probably of any scientific research in Great Britain".

In 1920 the Committee became known as the Medical Research Council and was granted a Royal Charter of Incorporation under the general direction of a committee of the Privy Council. A grant in aid was provided by the Treasury and has steadily increased until it now stands at some £14 million per annum.

Consider what happened at the bench. Until after the 1914–1918 War the pH meter for example (and indeed much electrical apparatus) was primitive and the electronics revolution had still to come. Analysts sometimes used colorimetric methods, the popular instrument being the Lovibond Tintometer with its boxful of calibrated strips of coloured glass. Discovery and characterization of a new compound meant obtaining enough material for analysis and the chemist was expected to carry out his own 'macro' C, H, N and S determinations as well as cryoscopic or ebullioscopic molecular weight determinations. Absorption spectroscopy, whether in the visible or ultra-violet region, was in the main an activity for specialists striving to build up a body of doctrine. Some very able organic chemists seemed at that time to regard recourse to spectroscopy as a sign of weakness—not far removed from cheating! The technique was photographic, distinctly tedious and only semi-quantitative. Advances in physics led in time to photo-electric spectrophotometry which made a great impact on chemistry and biochemistry. Physical chemists were refining methods in infra-red spectroscopy and the new techniques were to influence nearly all studies on natural products. The Warburg manometric methods which were both elegant and productive later met severe competition from the shrewdly exploited measurement of ultra-violet absorption at 340 nm. due to the reduced nicotinamide moiety of NADH or NADPH.

The simple centrifuges available before the first World War were successful enough to encourage the manufacturers to design and market better and bigger machines. Today the range of centrifuges, and particularly of high speed refrigerated centrifuges, meets a great variety of needs, and without such instruments the growth of biochemistry would have been stunted indeed.

Developments in microscopy have been very relevant to biochemistry, and electron microscopy after a rather slow start has now become so revealing as to be indispensable for many problems. Certainly many insights into the nature and structure of organelles are the direct outcome of electron microscopy, and correlated with biochemical observations the pictures are interpreted with imagination.

Biochemistry remains deeply involved in the discovery of new natural products as well as in purifying synthetic substances. The various kinds of chromatography — adsorption, ion-exchange, thin-layer and gas-liquid, together with preparative electrophoresis, have transformed the technique of isolation. And today the need to obtain relatively large amounts of material is less compelling than ever before. Microanalytic methods have become standard and by the time ultraviolet and infra-red spectra have been recorded automatically, the nuclear magnetic resonance spectrum examined and high resolution mass spectroscopy carried out, the constitutional problem has either been solved or at least greatly simplified.

In the separation and purification of large molecules similar advances have taken place and the elegant methods now in use for determining the structure of a protein would even 30 years ago have been unimaginable. Automation, notable in this field, is spreading into many segments of biochemistry including of course clinical biochemistry. Synthetic methods too have leapt forward.

The determination of structure in large and complicated molecules owes much to X-ray crystallography which in turn is indebted to developments in computer science. The advance in physics which led to ready availability of isotopes, the skill with which radio-active elements have been introduced into organic compounds and the advent of the newer counting techniques have deeply affected the planning and execution of biochemical research. If today the older biochemists pause, it is to rub their eyes with wonder.

B. H. Flowers (speaking as Chairman of the Science Research Council [*Chemistry & Industry*, 14 September 1968, p. 1233]) mentioned that 100 Mc/s nuclear magnetic resonance instruments cost £20,000 to £30,000 while the new 200 Mc/s instrument costs about £67,000 plus £5,000 per annum for liquid helium used in operating the superconducting magnet. So far as possible the new instrument would be shared between departments. In principle it could distinguish between different regions of a single protein polypeptide chain in aqueous solution and its performance was in the course of being evaluated. For £64,000 a computer could be linked to a mass spectrometer and used *inter alia* to calculate automatically the amino acid sequence of a polypeptide. Optical rotatory dispersion and circular dichroism could provide valuable information on the helical content of proteins.

Increased expenditure imposes its obligations on the scientific community. The system existing in this country has the great merit that working scientists, familiar with techniques, play a considerable part in the scrutiny of application to the Research Councils for grants. Large expenditure is indubitably necessary alike for training advanced students and research workers in the newest methods and for advancing knowledge. It should be said that many countries, including Britain, have profited also from unexampled generosity in respect of grants for research projects.

Enough has been said to show how the growth of biochemistry has coincided with impressive increase in instrumentation. This should not blind us to the equally striking intellectual stimulus brought about by ideas from other disciplines—botany, genetics, microbiology, physiology, pathology and histology. Reciprocal relationships between disciplines have without exception enriched both subjects.

In the second phase of the history of biochemistry (from say 1910–1950) it was the study of nutrition and of deficiency diseases in man and animals that provided the great challenge and also attracted some much needed financial support. Vitamins and hormones possessed a degree of glamour in offering to chemists and biochemists some chance of a personal contribution to the large-scale alleviation of suffering. The appeal was enhanced by the fact that at first the criterion for controlling chemical separations was biological—the animal test being indispensable. Biochemists (including at that time a distinctly mixed bag of recruits diversified in training and outlook) had to learn to carry out quantitative bio-assays. Moreover they had to enlist the help of statisticians and it was fortunate that statistical methods were in the throes of rapid development. Nutritional experiments provided good exercises and the biochemists were apt enough as pupils.

Over a rather long period the animal was often used crudely as a measuring device, and by the time isolation, characterization and synthesis had been achieved bio-assays were hopefully replaced by more speedy and more accurate physical or chemical methods. This did not mean that animal experimentation could be abandoned; deeper issues were at stake and there was need to link the biochemistry of deficiency states with physiology, histology and pathology. Those chemists and biochemists who took vitamins as the focus of their activities had the satisfaction of seeing some important modes of action explained in terms of coenzymes with specific functions central to metabolism. They also witnessed the early stages of progressive improvements in health attributable to the spread of knowledge concerning vitamins and hormones. They had a part in technological advances culminating in syntheses at economic prices and in the fortification of foods such as margarine and white bread. But of course many problems remained unsolved and in some cases serious lacunae persist to this day.

The rise of endocrinology was intellectually exciting and had vast consequences. The characterization of polypeptide and protein hormones and the isolation of the structurally simpler hormones, followed by elegant syntheses, marked a further advance deeply rewarding to medicine. By the synthesis and use of new biologically-active substances the Pharmacopoeias have been enriched. The total effects on the social scene have been immeasurably great. Many members of the Biochemical Society can look back on special occasions when they listened to announcements of important discoveries or achievements in the fields of vitamins and hormones. No doubt they went home in a whirl of admiration and hope often mingled with apprehension! In endocrinology as in vitamin studies rather large commercial operations became necessary and patents were inescapable. In the end issues of principle had to be decided by bodies like the Medical Research Council. However deeply scientific investigators might value calm and seclusion few biochemists could henceforth expect to retain all the privileges of an ivory tower. The lag between initial discovery and commercial production was shortened and collaboration was further encouraged by the dramatic achievements of chemotherapy and the impact of antibiotics. Closer contact with industry helped to stimulate intellectual exchanges internationally as well as nationally and biochemists began to feel part of a world-wide community built around a true discipline. That biochemistry had become a discipline in its own right was by this time widely accepted although some critics still regarded it as an untidy subject with ragged edges.

Perhaps, however, the most crucial change in biochemistry during the present century has been in elucidating metabolic sequences and in establishing the significance of high energy compounds such as ATP. The body of doctrine well supported by evidence is attested by the elegant metabolic charts which decorate the walls of laboratories. Admirable as summarizing great advances, these charts may be as useful to students as they are intimidating to their teachers! The picture includes fusing together almost a plethora of ideas and techniques, as in the beautiful but still unfinished work on mitochondria as sites for electron transport, the cytochromes and oxidative phosphorylation, and in elucidating the photosynthetic processes occurring in chloroplasts. Soon after World War I, when the quantum theory was new and half-developed, photochemistry and band spectra exercised a strong appeal. Chemists thought of synthesis in the green leaf as in the main a photochemical problem. It is now clear that the biochemical aspects were underestimated and the whole approach foredoomed by the inadequacy of the total scientific background. It was a gallant failure, but even after this disappointing experience a new generation of chemists and biochemists was often slow to change the approach.

New ideas brought about great advances; the photochemistry and the biochemistry were separated but the whole subject has turned out to be very complex. The paramount need to discover all the participating compounds has now been appreciated. Significant advances have doubtless been made in spite of gaps in basic facts, but just as an adequate performance of a stage play needs a full muster of all the players, it now appears certain from the extended roll-call of participating molecules that the study of electron transport and photosynthesis has suffered from missing links.

Biochemistry has always had a foot in the fermentation industries but it has now invaded the agricultural and food industries on a large scale. There have been great advances in the design and usage of pesticides, herbicides, preservatives, antioxidants and emulsifiers. Since few such changes are wholly advantageous new problems have arisen concerning safety-in-use and long-term dangers. Toxicological studies have been necessary and biochemists have become anxiously aware of teratogenicity and carcinogenicity. A sustained and very successful attack has been made on the metabolism of foreign substances by animals and to a lesser extent by plants. Links with the central core of biochemistry have been strengthened and a new era in pharmacological biochemistry has dawned.

At the heart of biochemistry there has also been a lively and ingenious attack on problems of biosynthesis. The studies on sugars, fats, carotenoids, sterols, hormones and vitamins, have revealed wide areas with underlying elements of unity. Allied to this has been steady advance in the chemistry of natural products, no area having greater ultimate significance than that of purines, pyrimidines, nucleosides, nucleotides and nucleic acids. This consolidated work opened the way to elucidating the significance of RNA and DNA, the double helix, coding and protein synthesis with all its implications and complications. Biochemists in general have accorded unstinted praise to the major insights and have quickly adapted themselves, both in teaching and research, to the new advances.

Throughout this period of astonishing progress biochemists have worked on micro-organisms, viruses, plants and animals. Comparative biochemistry has made great strides and such topics as ruminant digestion, growth, lactation and egg production, have been studied in detail. Just as the study of inborn errors of metabolism is an example of a growing subject resting on a firm base of interacting disciplines, so too the convincing work on the chemistry, biochemistry and physiology of vision illustrate the meeting of several streams of thought. Immunology, psychological medicine and resistance to infection are other fields where one discipline helps another. Avian biochemistry, insect biochemistry and fish biochemistry offer great opportunities for the future.

There can be no doubt that biochemistry is now a focal element in biological science. In this country it has crystallized a genuine group loyalty around and through the Biochemical Society. Furthermore the widening influence of the subject has broadened and intensified the Society's commitment. Systematic efforts have been made to foster international contacts between biochemists, to promote educational advances in diverse ways and to encourage social responsibility.

Almost inevitably there have been differences of opinion over special problems and groups and particularly over names. Thus biochemists have never taken kindly to the expression 'molecular biology'. In this they have resembled Ruskin in his approach to the Pre-Raphaelite Brotherhood, for although he repeatedly described the name as unfortunate, unwise and even ludicrous, he gave to the work greater understanding and more praise than any other instructed critic of his day. Biochemists have enormously admired—and imitated—the work which they regarded as a splendid blend of biophysics and biochemistry and perhaps fearing a beam in their own eye have decided after all,

"what's in a name? that which we call a rose
By any other name would smell as sweet".

Inevitable too was the formation of new societies, in nutrition, in endocrinology, in clinical science, in microbiology and other fields. The history of the Biochemical Society involves relations with many other bodies and institutions and however much opinions may differ on particular controversial matters it is certain that the total achievement has been considerable, thanks to a great deal of enthusiasm and hard work by many men and women.

THE LEGAL STATUS OF THE BIOCHEMICAL SOCIETY

In May 1925 a definite recommendation that the Society should be incorporated was made by its advisers, Thos. Gardner & Co., Chartered Accountants. The Committee then began to consider the matter, without it seems much sense of urgency. In April 1928 a subcommittee consisting of J. A. Gardner, A. Harden, P. Hartley and H. D. Kay examined the possibilities concerning the legal status of the Society. Rejecting (a) the idea of an application for a Royal Charter or (b) the formation of a limited company registered at Somerset House, they were left with a choice between incorporation under Section 20 of the Companies Act and a Trustee system. Section 20 was specially designed by the Board of Trade to meet cases such as learned societies wishing to have the status of a registered body but not desiring to be styled 'limited'. The subcommittee failed to agree and later in the year the main Committee decided by a narrow majority in favour of a trustee system.

The matter came up again in 1943 and as it appeared that there were advantages and disadvantages in incorporation no decision was taken pending a review of the general policy of the Society.

A new subcommittee (J. H. Bushill, E. C. Dodds, W. T. J. Morgan and F. G. Young and later A. C. Chibnall and W. Robson) was appointed in 1944. It reported that the chief advantages of incorporation were in limiting the financial responsibilities of officers and committee members; the Society was a collection of individuals whereas an incorporate body is an entity; it may hold property in its own name and it may sue and be sued. "An incorporate body is strictly bound by regulations, periodical returns must be made and there must be a registered office and a register of members open to inspection when required." The subcommittee reported in favour of incorporation but the main Committee deferred the issue for a further two years.

Many active members of the Society still preferred the Trustee system. H. Raistrick became a Trustee in 1946 and J. L. Baker, J. C. Drummond, C. R. Harington, J. H. Bushill and R. A. Peters all served at different times. In 1954 a supplementary Trust Deed vesting the Society's property in the Trustees of the Society was sealed.

The Minutes of the Committee record decisions on the functions of Trustees (1960) and confirmation of the then Trustees, (Bushill, Chibnall, Harington, Kay, Peters and Raistrick, 1961). It was agreed in 1962 that the Minutes of Committee meetings should be sent to Dr Bushill as convenor of the Trustees so that they might be "kept better informed of the Committee's decisions and activities", but at the next meeting it was decided to send the complete Minutes to all six Trustees.

In 1964 the Trustees became troubled about the consequences of a change in the Rules of the Society passed in 1960. "The Trustees were now clearly bound to act on the instructions of the Committee and felt some uneasiness in that they were accepting a measure of responsibility for funds over which they had no control in respect either of expenditure or investment policy". Harington, speaking for the Trustees, indicated to the Officers that the difficulty did not seem to be soluble within the 'present' constitution of the Society and suggested that the Committee might consider incorporation of the Society under the Companies Act. The Committee approved the Treasurer's action in making preliminary enquiries of the Society's legal advisers and approved expenditure on further consultations.

By September 1964 plans for incorporation had been approved by the Committee and were to be discussed at the Annual General Meeting.

In 1965 it was stated by Counsel that under the existing rules payment of honoraria (to editors, etc.) was illegal. The Committee thought amendment of the Rules was undesirable; Counsel's opinion only made Incorporation more urgent. It was then decided to print and circulate documents, with an explanatory note so that members could reach a decision at a special General Meeting to be held at Oxford on 15 July 1965. It was proposed that the Society should become incorporated as 'The Biochemical Society' and special approval would be sought from the Board of Trade to omit the word 'Limited' from the name.

Incorporation had become desirable because of the rapid growth of the Society and the financial turnover. In 1964/5 the revenue account was nearly £100,000 and the procedure by which the Trustees invested the Society's funds would be facilitated and contractual arrangements could be simplified.

A 'Memorandum and Articles of Association of the Biochemical Society' were prepared under the provisions of the Companies Act referring to a company limited by guarantee and not having a share capital. The Memorandum of Association detailed the various objects for which the Society is established and its methods of conducting its financial affairs. The Articles of Association were drawn up with great care and it seems unlikely that many occasions can arise for which adequate provision has not been made. There are nineteen sections covering membership and subscriptions, twenty-one sections dealing with organization, five with general meetings, ten with proceedings at general meetings, seven with votes of members, one with the seal, three with publications, four with accounts and two with audit, two with notices, one with dissolution and one with indemnity to officials.

THE LEGAL STATUS OF THE BIOCHEMICAL SOCIETY   35

This important landmark in the history of the Society was the result of a great deal of careful thought by the Committee and the Secretaries.

At the Oxford Meeting the proposals were approved and following completion of negotiations with the Board of Trade, the Society became incorporated under the Companies Act 1948 on 25 November 1966. As a result the role of the Trustees came to an end and the Report of the Secretaries (1966–67) took the opportunity to "thank all those who have acted for the Society in this capacity for their services over many years." On the advice of the Treasurer the Committee decide to set up a Finance Subcommittee to act in an advisory capacity. The Finance Subcommittee consists of the Honorary Treasurer (Chairman), the Honorary Committee Secretary, a representative of the Editorial Board of the *Biochemical Journal*, and three other members.

AIMS AND ADMINISTRATION OF THE SOCIETY

In 1921 the Committee decided not to revise the Constitution, but laid down that the Chairman of the Committee should be regarded for official purposes as the President. In 1958 the position of the Chairman of the Committee was discussed and it was agreed that it was highly desirable that he should be "less anonymous than he had been in the past", and in order to facilitate the Chairman's attendance at meetings, he should be entitled to have his expenses paid by the Society at the same rate as for the officers.

In 1936 it had become clear that on account of the rapid growth in size and activities of the Society two secretaries should be appointed, one to deal in the main with meetings and the other in the main with business of the Committee. This was agreed. An Honorary Assistant Treasurer was first appointed in 1941 so as to be able to take over in an emergency.

In 1942 limitations of period of service of Officers of the Society and Members of the Editorial Board were laid down.

A very important Committee meeting took place in March 1944 on the optimistic basis that the end of the war was coming. A very full discussion of the future policy of the Society occupied a whole morning. This was the last meeting for outgoing members of the Committee, but the decisions reached were unanimously approved by the new Committee, meeting at the end of the day. The resolutions were:

1. That the Committee welcomes the formation of new Societies with interests cognate to or overlapping, in some degree, those of the Biochemical Society but is strongly of the opinion that the objects and activities of these new organizations should be co-ordinated with those of existing Societies. Unless co-operation exists between different Societies with common interests much overlapping of activity will be inevitable.

2. That the Biochemical Society should initiate the formation of a Biological Council with functions similar to those of the Chemical Council: This would be best effected by informal approaches in the first instance to the Committee of the Pathological and Physiological Societies and to that of the Society for Experimental Biology, for a united movement on the part of four of the interested Societies.

3. That the activities of the Biological Council should be co-ordinated with those of existing similar organizations, such as the Chemical Council.

4. That the wishes of the members of the Biochemical Committee would best be met by a substantial increase in the number of discussion meetings arranged each year.

5. That developments in the teaching of biochemistry should be assisted by the Society where possible, in a manner to be determined as occasion may arise

K. S. Dodgson,
Honorary Committee Secretary

A. N. Davison,
Honorary Meetings Secretary

A. P. Mathias,
Honorary International Secretary

A. I. P. Henton,
Executive Secretary

In order to implement Resolution 2, R. A. Peters, F. G. Young and W. T. J. Morgan were requested by the Committee to make informal contact with the Societies named, and with regard to Resolution 4, the Committee formed a subcommittee to advise them on the matter.

More than 20 years were to elapse before the major policies of the Society were to be similarly discussed. In December 1965 the Committee set in motion a thorough scrutiny of future plans and policy. The accommodation problem was discussed in detail and several possible solutions were canvassed. It was said with some firmness that the Society had become more than the publisher of the *Biochemical Journal* and the body responsible for organizing meetings. It had need to concern itself about the relationship between biochemists working in industry and the Society. It was remarked that the extremely thorough way in which papers were edited tended to make them all read as if they were written by the same person and it was suggested that this tendency to uniformity could drive away some authors. One member went further and remarked, more in sadness than in criticism, that the *Journal* was getting dull. Perhaps there were 'good' papers to be got from Europe and perhaps the sense of dullness reflected the wide range of the papers with the result that few could be equally interesting to all members. Could not the meeting be divided into sections? Would it not be a good idea to set up groups? Need every paper on the agenda for the meeting be actually read? It was thought that the Agenda Papers, suitably enlarged, could themselves be a channel of communication. Certainly some new type of publication was needed to report meetings or less formal discussions.

Finally some consideration was given to what other new activities the Society might engage in. There should "emphatically be no extension into trade unionism", but there was general agreement that more should be done by way of collecting and disseminating information regarding vacancies, representing biochemists' views on such topics as financing research and fostering the relations between biochemistry and the medical sciences. This could be done by an extension of the discussion type of meeting such as that on 'Biochemistry in Industry' and by the production of a greater range of occasional publications. There was scope, too, for the Society's educational work: careers guidance at schools, more public lectures and teaching aids like *Essays in Biochemistry*, the collection of information on films and audiovisual aids, experiments with closed-circuit television in university departments, etc.

In June 1960 it had been decided that the Society needed a full-time administrative secretary to look after the day to day work of the Society and to assist the Honorary Secretaries. In December of that year Mr G. W. McHardy was offered the post at a salary comparable to that of a University Lecturer and with similar annual increments. Mr McHardy rendered very valuable service to the Society and on his resignation in 1966 he was succeeded by Mr A. I. P. Henton and the post was renamed that of Executive Secretary.

The 1965 meeting however was no more than a preliminary canter. A planning subcommittee was set up and reported towards the middle of 1966. It discussed the central administration and the transfer of much work from the Honorary Officers to the Executive Secretary.

The major expansion in the Society's activities was likely to be in publications and in supporting innovations in education. It was thought that the Society could act as a centre for information about teaching aids generally and might even have an advisory function on design of laboratories and the selection of equipment.

The main Committee accepted the idea of dividing Society meetings into colloquia, discussions and special lectures as well as communications and demonstrations. It supported the formation of groups. An interesting paragraph from the report of the meeting in July 1966 reads:

> "It seems of vital importance, in order to avoid 'splintering' of the Society, to encourage 'groups' (for example the Molecular Enzymology Group) to hold at least one of their meetings to coincide with an Ordinary Meeting of the Society. The possibility exists that more than one Discussion Forum or 'group' meeting could be held in different rooms at the one Meeting.
>
> "It may be important to take full advantage of all Meetings where hostel accommodation is available by ensuring that the Meeting extends for one or two days. On the other hand, most London meetings are held during term time when flexible arrangements are difficult to make. It might therefore be wise to continue to restrict most London meetings to one day and to make a point of including Lectures and Special Lectures as an additional early evening feature of such meetings."

The Committee was in favour of a variety of publications and thought that eventually there should be Proceedings of the Biochemical Society with a new subscription structure.

> "The Society should try and break down the division that existed between 'glamorous' and 'unglamorous' biochemistry by providing a milieu that was not so narrow that molecular biologists, for example, could not feel at home in it; the object should be to draw molecular biologists into the Society's membership. The Society should be more interested in biochemistry in industry, and the work done in Colleges of Technology. There was much that the Society could do on the educational front.
>
> "There was need for closer co-operation with the Chemical Society on the subject of meetings for the recent splitting of the *Journal of the Chemical Society* into 12 sections (two of which covered material accepted by the *Biochemical Journal*) showed how real was the danger of both Societies setting up groups of similar interests and of similar title even."

The Society, however, was not entirely preoccupied by big tasks, there were also watch-dogs active on smaller issues. In 1964 exception was taken to the informal way in which a retiring Honorary Secretary tended to find his own successor. This criticism was supported and the Committee decided that in future there should be a nominating committee consisting of the Chairman, the Honorary Officers and two ordinary members of the main Committee.

Another example of watchfulness occurred when in 1968 the Committee was challenged on the principle of appointing non-voting co-opted members of the Committee. After a brisk discussion it was admitted that the appointments were outside the terms of reference laid down in the Articles of Association, but it was agreed that the Committee should not be deprived of the advice of non-voting co-opted members pending further investigation.

The publications of the Society and its general business increased so much that the permanent staff was steadily enlarged. New activities, to be described in other sections, made heavy demands on officers and staff and the Society has gradually become a substantial business concerned with international as well as national activities.

W. J. Whelan's experiences as Meetings Secretary provide an indication of the amount of work facing the holder of that office. For 48 years notices of meetings had been printed in London but in 1959 following a printers' strike the work was transferred to Castle Cary in Somerset. The Meetings Secretary was often hard pressed to get abstracts off to the Cambridge University Press and notices of meetings to Castle Cary. "By setting aside practically a whole day to edit the Abstracts and prepare the announcements I could cut the overall time to a minimum but I found myself working against the clock and my wife became used to a regular midnight drive around London, first to put the Abstracts on the overnight train to Cambridge from Liverpool Street and over to Paddington to catch the corresponding train to Castle Cary for the announcements. Then about 10 days later came the proofs of the announcements and about 19 days after the press day the Abstracts would be delivered to Castle Cary for mailing with the announcements".

About that time the Secretaries persuaded the Committee to print the names of Committee members and officers on the inside page of the *Journal*. The presentation of printed matter was reorganized. A tidy octavo leaflet replaced miscellaneous pieces of paper, and later, announcements and abstracts were combined in quarto form. The abstracts, instead of appearing in order of receipt were grouped according to subject matter. The announcements of vacant research posts (first suggested by P. N. Campbell) were combined with notices of meetings in display sheet form. These innovations were widely appreciated.

PROCEEDINGS

In February 1924 the Committee decided that publication in *Chemistry and Industry* of proceedings of meetings together with short abstracts of the papers presented would satisfy the desire for prompt publication of communications. A proposal in 1926 to publish Proceedings in the *Biochemical Journal*, the cost being borne by the contributor at the current rate, was rejected (eight votes to two). Towards the end of 1926 advance galley proofs of abstracts which could be circulated at meetings were discussed but the Committee thought publication of abstracts in *Chemistry and Industry* would meet the need. This arrangement was apparently unchallenged for 15 years. In November 1941 the question was reopened of publishing Proceedings in the *Biochemical Journal* rather than in *Chemistry and Industry*. The Senior Editor was prepared to recommend to his Board that abstracts be published in the *Journal* "provided that a clear statement was made absolving the editors from responsibility for the content or method of presentation of the abstracts". In 1942 the Secretaries reported that the Cambridge University Press estimated a cost of about £7. 10. 0d. per meeting for the provision of pre-circulated abstracts and for binding the Proceedings in the *Journal* with a separate index. The increased expenditure was a small item in the total expenditure of the Society and the Committee approved the plan.

In March 1963 it was recognized that with continued expansion of the *Journal* the time would come when members could not be given the *Journal* without further payment over and above the annual subscription. Another publication, for example Proceedings, would then be desirable. It was thought that the Agenda Papers could be increased in size to match the size of the *Journal*. The stapling together of the Proceedings and the enlarged Agenda Papers was now approved in principle. This decision was put into effect and the result has given general satisfaction. All members are kept well informed of the Society's varied activities.

ACCOMMODATION OF THE SOCIETY

For many years the Society held its meetings at various colleges and medical schools in London and at other centres outside London, but it had no headquarters and no permanent office staff. Towards the end of the 1939–1945 war the general problem of finding suitable accommodation for British learned societies was rather widely discussed. It was foreseen that if the post-war period resulted in the expected expansion of scientific effort the Biochemical Society would need a headquarters in London; it could not indefinitely expect the administration of the Society's affairs to be conducted on a voluntary part-time basis by editors and officers using the

The Society's offices at 7 Warwick Court

AIMS AND ADMINISTRATION OF THE SOCIETY

The Senior members of the office staff, September 1969

Mr. S. W. New
Editorial Assistant

Miss Doris Herriott
Assistant Executive Secretary

Mr. A. G. J. Evans
Editorial Assistant

Mr. E. N. Maltby
Editorial Assistant

Mr. C. A. Sabner
Accountant

Mr. C. S. Wilson
Editorial Assistant

Dr. J. D. Killip
Editorial Secretary

Mrs. Mollie Bird
Senior Assistant, Editorial Office

premises of their employers. In 1950 the Committee Secretary represented the Society on a Scientific Societies Accommodation Committee set up by The Royal Society "to advise on the accommodation requirements of Scientific Societies in the proposed Science Centre and to make recommendations regarding construction and use of accommodation". The Minutes noted, however, that the centre "was not expected to be completed for some 10 years".

When E. J. King was Chairman of the Editorial Board the editorial office was situated in his Department at the Postgraduate Medical School at Hammersmith. In 1952 A. Neuberger took over the Chairmanship of the Board and the 'office' was transferred to the Medical Research Council's laboratories at Mill Hill, the Society paying £50 per annum rent to the M.R.C. "for two years plus telephone". Neuberger was authorized to spend £150 on furniture and up to £800 per annum for the services of a trained sub-editor. The move was made on 1 April 1952.

In 1954 informal discussions took place on the possibility of renting office accommodation from the Linnean Society at Burlington House, but by March 1955 Neuberger was telling the Committee that "finding alternative accommodation for the Editorial Office would become urgent in the near future". A subcommittee appointed to deal with the matter found that some rooms at the Lister Institute would shortly be available at a rent of £200 per annum plus £20 for library facilities. The accommodation would be available for two years at three months' notice. The Treasurer was authorized to complete the agreement with the Lister Institute and to open provisional negotiations with the British Medical Association with a view to a more permanent arrangement. Nothing came of this and a formal approach was made to the Principal of London University to see whether the Society could rent "1300–1500 square feet of accommodation (excluding landings and corridors) but including four rooms of total floor space at least 800 square feet". That was in 1956, but in 1958 the Editorial Office was still housed on the top floor of the Lister Institute and was using an extra room for the storage of records at a rent of £25 per annum.

In October 1959 the Director of the Lister Institute (Ashley Miles) wrote to say that the rooms now used by the Editorial staff were required for other purposes and asked for them to be vacated by Christmas or not later than March 1960. The Committee hoped that the Society could become a sub-tenant of the Medical Research Council in the new premises being constructed in Park Crescent and the idea looked very promising but there was a difficulty in that occupation could not be earlier than the summer of 1961. Temporary accommodation was therefore essential and suitable premises were found to be on offer at 133–135 Oxford Street at a rent of £675 per annum excluding rates. The tenancy was agreed but a seven year lease had to be signed. The Chairman of the Editorial Board was authorized to spend £300 on furniture. The staff moved into the Oxford Street offices in April 1960. Negotiations continued with the Medical Research Council concerning offices at Park Crescent on the basis that the sub-tenancy might last for seven years but could not be a permanent solution of the Society's accommodation problem. Other enquiries were conducted through The Royal Society.

In February 1961 an agreement between the Medical Research Council and the Society was in preparation by the Treasury Solicitors. The total rent was £1425 per annum. In April temporary accommodation for the Administrative Secretary (Mr McHardy) was found in the Biochemistry Department at University College London. The move to Park Crescent took place in September 1961 and by December 1961 a sub-tenant had been found for the premises at Oxford Street on terms favourable to the Society.

But the Society's accommodation problems were not over. Opinion had slowly crystallized that the Society would be well advised to acquire its own permanent headquarters, but nobody doubted this would be costly and that reserve funds would have to be mobilized. H. R. V. Arnstein recalls that on one occasion a special subcommittee lasted from early evening until 4 a.m. After much discussion a Finance Subcommittee to advise the Treasurer on investments and general financial policy and to make recommendations to the Committee of the Society was set up in 1966. The first meeting was held in December of that year. It was decided, with a view to purchase, to make enquiries concerning suitable premises appropriately located.

It did not prove at all easy to find what was needed but eventually a promising property was considered. The Finance Subcommittee obtained professional advice and subjected the proposal to detailed examination but had to reject it. Fortunately another building at 7 Warwick Court, London, W.C.1 was discovered and the main Committee was able to endorse a recommendation from the Finance Subcommittee to purchase the premises for £57,094. A small part of the house was later let at a reasonable rent.

Funds for the purchase of 7 Warwick Court were raised from the Society's reserves by the sale of about 70 per cent of the investments. The purchase of the new headquarters was the largest transaction the Society had ever undertaken. The need was so great that the decision had to be taken despite the fact that the time was not very favourable for the realization of investments.

The purchase of 7 Warwick Court placed a heavy responsibility on those concerned and the Society is much indebted to them for the wisdom of their actions and the time they gave to the task.

FINANCIAL AFFAIRS

R. H. A. Plimmer acted as both Honorary Secretary and Treasurer for the Biochemical Club, but when the Biochemical Society was properly constituted in 1913, J. A. Gardner became the first Honorary Treasurer. He saw from the outset that the Society would need to build up reserves not only for 'the rainy day' but to permit uninterrupted growth—particularly of the *Journal*. Gardner, according to Plimmer, realized very early that sooner or later editorial services would have to be paid for.

Gardner held the Treasurership for 31 years and only in his last year did he ask for the help of an Assistant Honorary Treasurer. Even then he seemed to have been mainly influenced by the risk that the consequences of enemy action might be more serious if all the information about the Society's finances was in one person's hands.

In 1941 a generous offer of assistance had come from the United States. C. G. King suggested that it might be useful if the editorial office of the *Journal of Biological Chemistry* served as a filing agency or repository, for the protection of scientific records and complete or incomplete manuscripts that might later become available for publication.

About the same time in 1941 the Committee's reaction to air-raids was the appointment of an Honorary Assistant Treasurer "to provide that somebody was able officially and easily to take over the full duties of the Treasurer at short notice should such a course become necessary during the present state of emergency. The Assistant Treasurer would be expected also to act as a general assistant with the ordinary work of the Society. It was also thought that under present conditions it might be wise to duplicate all the Society's records if that would be possible". J. H. Bushill was appointed and later succeeded to the Honorary Treasurership. Bushill has recalled that Gardner's stock answer to those who commented on the difficulty of prising money out of him was that "the job of a Treasurer is to treasure". Bushill himself continued the policy of building up reserves. He felt "that the ultimate success of the Society depended not only upon the scientific content of its meetings and the excellence of its *Journal* but also upon adequate financial backing to support a forward-looking and virile organization".

The finances of the Society had not always been healthy. Small profits came from the sale of the *Journal* for the first few years. After the first World War costs of publication rose and exceeded the available funds. Accumulated profits had to be used and it was only with the help of The Royal Society and a few gifts that Gardner managed to keep the

Society solvent. It became necessary to increase the subscription to 35s. per annum and to increase the cost of the *Journal* to institutional subscribers. The Cambridge University Press was very helpful and conceded a new agreement more advantageous to the Society. Nevertheless with the cost of printing rising steadily the small profits for the next few years proved inadequate and losses again ensued. Another grant from the funds administered by The Royal Society and a further reduction in charges by the Cambridge University Press helped to restore the balance. The subscription was raised to £2. 2. 0d. and outside subscribers paid £3. 10. 0d. for the *Journal*.

The turning point came with a sharp increase in membership and in sales of the *Journal* to institutional libraries. The Society then began to make profits and build up its reserves. Bushill commented as follows. "When the annual financial reports were presented it was not unusual for someone to point a finger at the surplus of income over expenditure and say that the accumulation of money was not the function of the Society. Some attempts, which deceived no one, were made at hiding the surplus by transferring money to a 'contingency' account and to justify such action by drawing attention to the continuing rise in the cost of publishing the *Journal*. It was emphasized that, with the increasing size and activities of the Society, paid secretarial and clerical assistance would some day be needed. That was a serious contingency against which, it was stressed, the Society must be prepared". Accordingly the reserves grew. The faction which criticised the accumulation of funds had its way over the price of symposium reports. As up-to-date summaries the reports had a special appeal and were published as cheaply as possible. But as the Treasurer noted "it required inspired crystal gazing to decide upon the number to be printed in order that costs and receipts should balance."

In 1947 Bushill also acted as Treasurer for the Chemical Council, deputizing for Professor Alexander Findlay who was in India. Dr Roche Lynch, having with the help of Messrs C. F. Chance & Co. revised the investment policy of the Society of Public Analysts, advised the Chemical Council to do likewise. Impressed by both operations Bushill recommended that the Biochemical Society should also seek the advice of Messrs Chance and act upon it. This was agreed and the portfolio was considerably changed. Messrs Chance have continued to advise the Society.

Bushill found that the administration of the Society's financial affairs was no sinecure, particularly in war-time. He induced Mr Henry Mears of J. Lyons & Co. to assist him in simplifying the Society's accounting system and when the Treasurer completed his term of office the Committee was pleased to present a gold watch to Mr Mears in recognition of his substantial services.

F. A. Robinson after a preliminary year of co-operation took over the Treasurership in April 1952 and the Committee agreed that he should be allowed part-time help from Mr A. N. I. Mann, an accountant on the staff of Messrs Allen and Hanburys.

In 1951/52 the financial situation again looked somewhat threatening and Robinson felt that the subscription should be increased to £3. 10. 0d. per annum. This was agreed to at a General Meeting held in November 1952 but at the Annual General Meeting held in March 1953 the decision was criticised by A. L. Bacharach and some other influential members, mainly on the ground that the Society had by then built up good reserves. The decision was however upheld. Robinson records "In fact we only lost 97 members as a result of the increased subscription and we had a credit balance of £1000 at the end of the financial year in 1954. I estimated that had we not increased the subscription we would have had a deficit of £1800."

In 1960 the Trustees obtained legal advice as to any restrictions which governed the way the Society's reserves were invested. Actually there were no restrictions when power of investment lay with the Committee and not with the Trustees. The Society's solicitors suggested an amendment to one of the Rules making it clear that the Trustees should "deal with the same as directed by the Committee". In April 1960 one-third of the Society's gilt edged securities were sold and shares in twenty different industrial equities were bought. It became necessary, however, for the Treasurer to discuss their position with the Trustees and the Committee agreed to invite the Trustees to accept responsibility for the investments on their behalf. It was also agreed to employ a firm of investment advisers. In 1963 it was arranged to have the investment portfolio scrutinized annually by the Society's stockbrokers. The solicitors advised the Society that the six Trustees who were all required to sign every document might agree to delegate power to sign to two of their members.

It will be seen that the prevailing conviction that the inflationary process required a transfer of some of the Society's assets to equities had resulted in appropriate action. It was however also clear that the Trustees were beginning to regard their position as a little anomalous.

In 1962 a special vote of thanks was recorded to F. A. Robinson who had been the Society's Honorary Treasurer for 11 years. He was succeeded by W. F. J. Cuthbertson who is still in office.

The *Journal* was still increasing in size and the activities of the Society growing. Inflation was still taking place. Ordinary membership of the Society in 1967 was £5 per annum or $15, student membership £3. 10. 0d. and a joint husband and wife membership was fixed at £7. 10. 0d. or $22.50. Nevertheless it was becoming clear that every new member who received the *Journal* automatically was financially a liability. In 1966 the run-on cost of the *Journal* was over £4 and the general activities of the Society also exceeded £4 so that each member cost the Society about £8. 10. 0d. per annum. The question naturally arose whether there should be a category of membership not accompanied by the right to receive the *Journal*. In October 1966 it appeared that a new member was a liability to the extent of £3. 15. 0d. per annum.

After thorough enquiry and discussion it was decided that membership only should be fixed at an annual subscription of £3. 10. 0d. and membership with the *Journal* at £9. 0. 0d. per annum. Any member would be entitled to receive *Clinical Science* for an additional £3. 10. 0d. per annum.

The new order resulted in a relatively small number (161) of resignations from the Society. The Membership in 1954 was 2,160, in 1961 3,000 and in 1969 had reached 4,620. About two-thirds of the members opted for membership only but the membership continued to rise and at the Annual General Meeting in April 1968 the Honorary Treasurer was able to report that some 1800 members were taking the *Journal*.

Dr Cuthbertson's period as Honorary Treasurer has thus included a very major change in the nature of the Society's relationship to its members. The publication boom, sometimes referred to as an explosion, sometimes as an avalanche, made such a change inevitable. It has been carried through remarkably smoothly. The same period also saw the purchase of 7 Warwick Court as the Headquarters of the Society, a transaction which because of its magnitude as well as its importance, must also be regarded as a major event in the history of the Society (see p. 47).

W. F. J. Cuthbertson, Honorary Treasurer 1962—

GROUPS

In July 1964 the Committee considered a proposal by H. Gutfreund to set up a *Molecular Enzymology Group* as a 'section' of the Society to hold eight meetings a year, mostly in London. The Society was asked to vote £100 for the purpose. The proposal raised important issues of policy on which the Committee proved very divided. The matter was placed on the agenda for a General Meeting in December. The Minutes for November 1964 record that a letter had been received objecting in this context to 'molecular' as a vogue and the Committee decided in favour of 'Enzymology Group'. The General Meeting held on 11 December approved a recommendation that the *Molecular Enzymology Group* be accepted as an official Group within the Society. In spite of the Committee's decision the objection to 'molecular' escaped attention. Shortly afterwards Rules were drawn up and approved for the conduct of the Group's affairs.

It soon became evident that the new Group was meeting a need and its success gave encouragement to the formation of other Groups. In February 1966, D. V. Parke invited the Committee to set up a *Biochemical Pharmacology and Toxicology Group*. Further information was sought and in April the proposal was discussed in committee, the outcome being left uncertain because of a reluctance "to undertake any considerable financial commitment".

Meanwhile a somewhat different situation had arisen over the possibility of forming a branch of the Society in Ireland as an alternative to forming a new Society.

This was not altogether a new problem. W. O. Kermack recalls a meeting of the Society in Edinburgh in 1921 when G. Barger rounded up the younger people to attend and to join the Society. By 1936 there was "considerable agitation to set up some sort of Scottish Association which would organize biochemical meetings in the Scottish University centres and give biochemists in Scotland more opportunities for meeting each other. . . . "It was in general not possible at that time for members of University staff or other employees to recover their travelling expenses to attend Society meetings and for Scottish members a visit to London could be quite expensive. Though railway fares were much lower, salaries were proportionally even more exiguous. I believe that it was not until during or after the 1939 War that even the Committee members were able to claim their fares. It is not surprising that there was a strong movement in favour of holding local meetings. But clearly the formation of a separate Biochemical Society was not the best way of solving the problem and accordingly many of us welcomed the compromise which was reached. This consisted in a gentleman's agreement that at least one meeting of the Society should be held in Scotland each year. They were to be in the Universities of Edinburgh, Glasgow, Aberdeen and St Andrews in rotation, the St Andrews meeting being alternately in St Andrews and Dundee so that we met, for example, in St Andrews every eighth year. As far as I know this arrangement was implemented without interruption until 1966. In 1967, the committee, evidently ignorant of the agreement (they knew not Joseph) arranged a meeting in St Andrews instead of Aberdeen, where the last meeting had been in 1963. They are making amends by meeting this year in Aberdeen, but of course one must recognize that times have changed and the biochemical centres in the provinces have much multiplied in numbers, so that the arrangements made in the thirties are now out of date and require the revision which I believe is now being undertaken."

In October 1966 a meeting of over 200 biochemists in Ireland decided not to form a separate Society. Instead they proposed an annual meeting in Ireland as against the 'present' arrangements of one meeting every six years. The Treasurer reminded the Committee that at present rates an Irish meeting would cost the Society about £700.

A subcommittee consisting of A. T. James, J. R. Knowles, D. C. Phillips and W. J. Whelan, having been appointed to look into the principles concerning the question of the establishment of Groups, met in October 1966 and produced a forthright report. The subcommittee came out strongly in favour of the formation of subject groups. In addition to Molecular Enzymology and Pharmacology the following list was put forward for discussion:

(1) Carbohydrates
(2) Steroids
(3) Proteins
(4) Nucleic Acids
(5) Intermediary Metabolism
(6) Immunology
(7) Bacterial Genetics
(8) Lipids
(9) Physical Methodology or Physical Biochemistry
(10) With less certainty, Industrial Biochemistry.

The subcommittee felt that the responsibility of the Society for such groups would lie in formulating general procedures to be followed. The initial action should lie with workers in the particular fields. The following procedure was suggested:

(1) At least twelve people should commit themselves in writing to supporting any proposed group. The signed proposal should be submitted to the Biochemical Society Committee for its consideration. The proposal should include a suggested organizing committee and also a defined field of operation.

(2) The meetings of such a group should be controlled by its organizing committee.

(3) Where the Biochemical Society Committee felt it appropriate, a grant should be given to the group to *partially* finance its operation.

(4) Secretaries of groups should be encouraged to approach the Meetings Secretary of the Society to investigate the possibility of occasionally taking over an ordinary meeting of the Society (including the normal financial arrangements for such a meeting).

(5) An annual meeting of all Group Secretaries with the Meetings Secretary of the Society should be held to co-ordinate meetings and to collaborate in joint meetings.

The subcommittee went on to outline proposals and procedures for collaboration with other Societies and it also reported on regional groups. It was felt that the latter would be useful when geographical isolation was a factor but not otherwise.

The main Committee had a very full discussion of the subcommittee's report and one fact at least emerged as generally agreed, namely, that there were issues of policy "that could only be settled at an Annual General Meeting".

In February 1967 a new situation arose over a proposal to form a joint *Carbohydrate Group* with the Chemical Society. The sponsors wished the arrangements to serve as a model for establishing further joint groups. Membership was to be free to members of both Societies and membership of the Group Committee should be divided equally between the two Societies. Financial responsibility should be divided equally between the two Societies and periodic reports concerning the Group should be rendered to both Societies. At the time of writing, however, the project for the establishment of this group had not come to fruition.

In February 1967 the Committee also agreed that a proposal to form a *Neurochemical Group* be submitted to the Annual General Meeting. Meanwhile the Group subcommittee had got its teeth into the problem. It recommended that a working party of the Biochemical Society should be set up to draw up concrete proposals for collaboration between the learned societies [concerned with chemistry and biology] similar to that started with the Chemical Society. It was also suggested that there should be a European equivalent to the American Gordon Research Conferences (A. T. James, H. L. Kornberg and P. N. Campbell were asked to constitute the Working Party).

The next proposal was to form a *Lipid Group* and in June 1967 a private meeting was held at Colworth House (Unilever Limited). Four papers were read and the 53 persons attending elected a temporary committee to arrange an inaugural meeting.

On 13 July 1967 the Society, in General Meeting at Oxford, approved the policy of forming groups and the *Irish Area Section* and the *Neurochemical Group* were established and joined the *Molecular Enzymology Group* and *Pharmacological Biochemistry Group* as legitimate offspring of the Society. In passing it may be mentioned that the *Irish Area Section* is experimenting in an interesting way with the reading of papers by predoctoral research students.

These were followed by the official recognition of the *Lipid Group* in February 1968 and the establishment of the *Immunoglobulin Discussion Group* in April 1968. A joint *Group for Steroid Biochemistry* formed between the Biochemical Society and the Society for Endocrinology won official recognition in February 1969 and in doing so became the first joint group of the Society. An embryo of a *Joint Computer and Instrumentation Group* formed in collaboration with the British Biophysical Society first met in March 1968, and was finally constituted in April 1969.

In July 1968 the subcommittee on groups pointed out that the organization of group activities had considerable implications, "even financial ones". There was a need for a close scrutiny of proposals to form new groups and amalgamation was recommended whenever it became desirable, but it is clear from the Annual Reports published in the Society's Yearbook that the various Groups are proving viable and have got off to a good start.

In recent years enthusiastic and determined members have succeeded in a great variety of ways in changing the character of the Society. Many of the changes were necessary and desirable and indeed some overdue. Experience has shown that a large proportion of the membership will neither oppose nor support 'reform' and votes taken at the end of an Annual General Meeting with a rapidly dwindling audience have settled controversial issues in a constitutional manner without generating much satisfaction. Indeed dynamic leadership is better sustained over major issues by the referendum than by a vote taken late at an Annual General Meeting. This is not so much a view personal to the writer as a summary of opinions volunteered to him in the course of preparing this History.

THE BIOCHEMICAL JOURNAL

In the early history of the *Journal* many revealing incidents, large and small, are recorded in the Minutes of the Committee. For instance, in November 1920 Dr Harriette Chick, then engaged with Miss Eleanor Hume on her classical studies of malnutrition in Vienna, wrote appealing for a free copy of the *Journal* to be sent to the University of Vienna "in view of the difficulty experienced in obtaining such periodicals in Central Europe". This was agreed for 1921. A very different note records that in March 1921 the Cambridge University Press reported a clerical error by which the Society had been credited at the end of 1919 with a profit of £135 instead of a profit of £5. The Society's own auditors had discovered the same mistake!

In 1923 a new agreement was entered into with the Cambridge University Press (C.U.P.). The Press would charge $12\frac{1}{2}$ per cent on the printing costs and $12\frac{1}{2}$ per cent on sales, the Society receiving the profits and fixing the price of the *Journal* and of reprints. In view of the additional responsibilities assumed by the Society under the new agreement a permanent subcommittee consisting of the Chairman, Treasurer, Secretary and Editors was set up to keep an eye on developments.

It was reported in February 1925 that the *Journal* had expanded by 500 pages (Vol. XVIII) and that a loss of £33 had been incurred on one year's working. A grant-in-aid of £100 was made by The Royal Society from the funds of the Government Grant Committee.

In 1929 a subcommittee (R. A. Peters, J. A. Gardner, A. Harden, C. R. Harington and R. Robinson) explored possible ways of reducing the cost of producing the *Journal*. They considered a new arrangement with the C.U.P. whereby the Press "would increase the rebate of $7\frac{1}{2}$ per cent on the printing charge to $12\frac{1}{2}$ per cent and waive the $12\frac{1}{2}$ per cent commission on members' copies". The change if made retrospective for 1928 would for that year save £243, about 10 per cent of the total cost of the *Journal*. The subcommittee recommended this new arrangement.

For some years Harden as Senior Editor had very valuable assistance from H. W. Dudley but in 1930 the latter found that editorial work was interfering too much with his research and felt compelled to resign. The Committee, still thinking that two men should be able to edit the *Journal* in their spare time, invited C. R. Harington (who had been one of the Secretaries for a year) to join Harden. Harington accepted and served from 1930–42 becoming Senior Editor in 1937.

D. G. Walker, Chairman of the Editorial Board of the *Biochemical Journal* 1969–

The expected deficit for 1931 due to the rising cost of the *Journal* was £400–£500. The C.U.P. offered additional minor concessions and the question of raising subscriptions was discussed by the Committee but "in view of the general financial position of the country, and especially of the Society members who might suffer salary reductions during the coming year" the Committee desired to keep the increase as small as possible. After much debate an increase in membership rates to £2. 2. 0d. was recommended and the subscription rate to £3. 3. 0d.

It was reported in 1933 that 23 per cent of the papers (and 27 per cent of the space) in the *Journal* came from non-members. Harington warned the Committee that a rule insisting on membership of the Society as a qualification for acceptance of a paper would have a bad effect. A very cautious line was taken on this issue.

In May 1934 it was decided that as from January 1935 the *Journal* should be published monthly instead of bimonthly and that a third editor would be needed. F. J. W. Roughton was appointed.

At the beginning of 1937, Harden, then assisted by Harington and Roughton, was still Senior Editor having held office since 1912. The Society had in 1929 acknowledged Harden's service by a gift of 200 guineas and had voted an annual honorarium of £100 from then onwards. It was decided in November 1937 to present Sir Arthur Harden with a silver salver, "suitably inscribed and with reproductions of the signatures of all those now living who have served on the Committee during the 25 years that he has been Editor of the *Journal*." The inscription read:

> To Sir Arthur Harden, F.R.S.
>
> From the members of The Biochemical Society. To express their appreciation of the inestimable services rendered by him to the Society as Editor of the Biochemical Journal from 1913 to 1937, and their gratitude for his friendship and wise counsel, available for all members at all times.

In June 1938, the Committee unanimously agreed to "invite Sir Arthur Harden to accept the nomination of the Society for election to Honorary Membership".

Sir Charles Harington has very kindly written the following notes on his period of editorship.

> "As it happened I welcomed the invitation, little realizing what I was letting myself in for, because at that particular juncture I had no serious responsibilities outside my own research and I was anxious for a task that required a different type of effort; this I certainly got. I had had, of course, no previous experience of editorial work and my appointment was an indication of the somewhat lighthearted view that the Committee at that time took of the duties required of the editors of the *Journal*. I am sure that the very thought of a professional editor would have filled them with horror.

By the time I joined Harden he had trained himself to be an excellent editor. He possessed an equable temperament, could work rapidly with economy of effort and was an admirable colleague for whom I had a great respect, which increased as time went on. I soon learnt, however, that he expected his co-editor to possess the same capacity for getting through the work as he himself had acquired; no sooner had I been appointed than he told me that he had arranged his summer holiday for certain dates which would mean that he would have to leave me to prepare the next number of the Journal for press by myself. I neither relished the prospect, which was somewhat alarming, nor enjoyed the performance—especially as this involved the almost complete re-writing of one of the papers—but there is no doubt that this drastic introduction did give me a measure of confidence (perhaps too much) and taught me in three weeks of hard work what I might otherwise have taken a long time to learn.

In the early and amateurish period of which I am writing, editorial practice was admittedly dictatorial. We did not expect our decisions to be questioned, nor did this often happen. We made little or no use of external referees, trusting our own judgement even in fields in which we could not really claim to be expert. The simplicity of the arrangements had the great advantage of avoiding delay and we took pride in being able to offer a speed of publication which I believe compared favourably with that of any other scientific journal of comparable standing. In this we were greatly helped by the speed and efficiency of our publishers, the Cambridge University Press. On the other hand, the lack of any assistance apart from minimal secretarial help did place a considerable burden of routine work on the editors; for example, we read all proofs ourselves, both galley and page, and from this task there could be no let-up during holidays or at any other time, if our reputation for prompt publication were to be maintained.

Scientifically we undoubtedly took risks in relying so completely on our own judgement, and I am sure that we must have made mistakes. Indeed, I remember two scrapes that I got into myself, one of which caused the resignation from the membership of the Society of a senior continental professor who took exception to an editorial alteration that I had made to one of his papers (fortunately he later returned to the fold); the other occurred when I referred back a paper by a senior biochemist in this country, and as a result had the whole of his department up in arms against me; here again, as it turned out, personal relationships were not permanently impaired.

Nevertheless, incidents of this kind were warnings of the more serious results that might ensue from editorial misjudgement, and at the same time the likelihood of such misjudgement was rapidly increasing owing to the rising flow of papers for publication and the broadening of the subject matter. For this reason Harden and I persuaded the Committee to allow us to recruit more editorial colleagues. We naturally sought for men who were expert in the fields with which we ourselves were less familiar and we were fortunate in obtaining the help first of all of F.J.W. Roughton to deal with papers involving physics and physical chemistry and later of S. J. Cowell and Frank Dickens to cover the fields of nutrition and of cellular biochemistry respectively.

With these accessions we were able to carry on reasonably well for a few more years, but there still remained the problems of proof-reading and indexing with which we had no assistance and which were becoming more burdensome with the continuing increase in the flow of material. In 1942 I was appointed Director of the National Institute for Medical Research and had perforce to give up my editorship; this afforded the opportunity for the Committee to consider how they wished the *Journal* to be conducted in the future. The decision was made to appoint an enlarged editorial board, and at the same time to introduce certain changes of policy, among which the most important was the use of external referees to help in the assessment of papers for publication as a matter of routine rather than as a procedure reserved for specially difficult cases.

These changes were the beginning of the development of the substantial organization that the Society now employs for the production of the *Journal*. The changes were inevitable and were probably overdue. They did, however, come in time to enable the *Journal* to keep pace with the enormous increase in biochemical research that has occurred during the past twenty-five years and to strengthen its position as one of the leading scientific journals of the world. That this should be the outcome is a more than adequate reward to those members of the Society who did their best to maintain the standards of the *Journal* so long as the task remained within the scope of amateurs."

The report of the Honorary Secretaries for 1942 to 1943 includes the following:

"Since the last Annual General Meeting, Professor C. R. Harington has retired from the post of Senior Editor of The Biochemical Journal. It is fitting, therefore, that we should recall the fact that Professor Harington was an Editor of The Biochemical Journal for 12 years, for the last 4 of which he occupied the position of Senior Editor with chief responsibility for the production and editing of the Journal. During this period a substantial part of his time and energy was devoted to this work, and, on behalf of the Committee, we wish to record a deep sense of gratitude to Professor Harington for his unselfish devotion to the interests of the Society over so long a period."

A special subcommittee (J. H. Bushill, F. G. Young, N. W. Pirie, W. T. J. Morgan, B. C. J. G. Knight and J. C. Drummond) was appointed "to consider the future of the *Biochemical Journal* in view of the impending resignation of the present Senior Editor (May 1942)". The subcommittee recommended "that the *Journal* be conducted by an Editoral Board of about six members, one to be appointed Chairman. Each member of the Board would be responsible for considering papers within given fields and for suggesting external referees when necessary. The Chairman would allocate papers". Honoraria were to be abolished but effective secretarial assistance provided for the Chairman. The subcommittee outlined appropriate sub-divisions of the biochemical field and suggested names of possible editors. It was also recommended that the Chairman and one member of the Board should be *ex officio* members of the Society's Committee. Members of the Board other than the Chairman should serve on the Committee for one year in rotation. It was also recommended that a volume of the *Journal* should contain not more than 600 pages. It was also agreed that names of authors should "give surname and initials only; no titles and no appendages".

The main Committee considered the Report (Dickens and Knight having left the room). The terms Senior Editor and Associate Editor were respectively to be replaced by Chairman of the Editorial Board and Member of the Editorial Board. In correspondence with authors of papers the Chairman should write on behalf of the Board as a whole and the Committee agreed to provide secretarial assistance for the Chairman. F. G. Young was invited to become Chairman and agreed to serve and a number of nominations to serve on the Board were made.

As the war went on pressure on space in the *Journal* slackened, delays occurred and there was a national paper shortage. This resulted in an excess of income over expenditure and the Committee prudently decided "that the sum of £1000 be set aside as a further contribution to the reserve fund initiated in March 1942". In May 1946 E. J. King succeeded F. G. Young as Chairman of the Editorial Board. At a luncheon in November 1946 tributes were paid to the valuable work done by Young.

An Honorarium of £200 per annum was offered to the Chairman of the Editorial Board and in 1950 the members of the Board were each voted £50 per annum. E. J. King was succeeded by A. Neuberger who held office until 1954, when he resigned on leaving the National Institute for Medical Research. The policy issue of appointing a full time Editor was then fully discussed. The Board and the Committee agreed in recommending (November 1954) that

> "the Chairman of the Editorial Board should have the help of a full-time Editorial Assistant, that this paid assistant should preferably be a biochemist or an organic chemist with research experience, possibly already retired from his official duties; that the initial salary offered should be in the range £700–£1200 depending on qualifications and that following this reorganization, the Chairman should have the same responsibility as he had had in the past for the work of the Board, and that the duties of members of the Board should remain substantially unchanged".

F. Clark was appointed Secretary to the Editorial Board in January 1955 and very ably held office until his tragic death in a motor accident at the end of 1968. Neuberger was succeeded by A. G. Ogston as Chairman of the Board later in 1955.

In 1952 the Minutes indicated some unrest about arrangements for printing the *Journal* and the Finance Subcommittee was instructed to enquire into probable costs of producing the *Journal* "by printers other than the Cambridge University Press" but nothing came of the matter at the time. The next stage was reached at the end of 1955 when the Committee asked the new Chairman of the Editorial Board to consider the possibility that the Society might become its own Publisher "in the light of the experience of The Royal Society in taking over the publication of its journals from the Cambridge University Press." Accordingly, A. G. Ogston, accompanied by T. S. Work and F. Clark visited Burlington House in January 1956 for discussions with Dr D. C. Martin and Mr J. H. Boreham. It transpired that The Royal Society had acted as its own publisher since 1954 and that net savings had resulted. It was concluded that a similar change-over would probably pay the Biochemical Society, in the long run if not at once, but that considerable problems of re-organization would arise. The Honorary Treasurer's reaction to Ogston's report was "that this matter should seriously be discussed at a future committee meeting but it is one into which we ought not to embark without adequate thought".

In October 1956 the Committee considered a letter from the Pergamon Press suggesting that the Society might care to open exploratory discussions with them with a view to their publishing the *Biochemical Journal*. The Treasurer and Ogston (then Chairman of the Editorial Board) were authorized to enter into such discussions and obtain a firm estimate for producing the projected Index to 25 Volumes as well as an appropriate estimate for producing the *Biochemical Journal*. These discussions did not however proceed very far.

By the end of 1956 the number of copies of the *Journal* to be printed had reached 5750 and it was agreed that the *Journal* "should publish papers in all fields of Biochemistry, plant, animal and microbiological, provided that the results make a new contribution to biochemical knowledge; or that they describe methods applicable to biochemical problems". The circulation of the *Journal* was increasing steadily and in December 1958 it was decided to increase the print order to 6000 copies. The growth in sales and in the size of the *Journal* underlined the need to keep a sharp eye on publication costs.

The Treasurer had reported early in 1961 that the Cambridge University Press proposed a new financial arrangement to be operated from 1 January that year. The commission of 15 per cent on all sales of the *Journal* and other publications would remain as previously. The commission on cost of production of the *Journal* had been $12\frac{1}{2}$ per cent but the proposed new basis was "a commission of $3\frac{1}{2}$ per cent of the volume price (or where there was no volume price the aggregate prices of the parts) multiplied by the number of copies of the volume being printed". It was calculated that on the figures for 1959 the Society would have saved about £900. The Treasurer was instructed to look into the effect of the new proposals and, after analysing the figures for 1958, 1959 and 1960, and extrapolating to 1962, he reported that the financial trend of the proposals was unfavourable to the Society. He would have preferred a sliding scale based on the old system whereby the percentage commission could be progressively reduced as the circulation increased—he hoped to 8000 copies. Later in 1961 the Minutes reveal renewed dissatisfaction over negotiations with the Cambridge University Press.

In March 1962 proposals were received from the Chemical Society regarding the distribution and publication of the *Biochemical Journal*. The Chemical Society was willing to act as distributor and it was calculated that 20 per cent of its activities in selling journals would allow the *Biochemical Journal* to be distributed for £2,700 plus about £250 for the storage of back numbers. The Committee became convinced that if the Society were to act as its own publisher and have the Chemical Society as its distributing agent, very substantial economies would result.

The Secretaries and the Treasurer were authorized "to broach the proposals to the Cambridge University Press and to report back to the Committee without in any way committing the Society". Once the matter had been raised with the Press it would be in order for the Editorial Board to be given copies of the proposals and to assess them at a meeting. Some members of the Committee were loath to break with the Cambridge University Press; "the Society's relations with the Press were of long duration and had been most harmonious".

It soon emerged that the Press would decline to print the *Journal* if it did not also publish it. This meant that if the Society were to go ahead with the scheme it would have to find printers and seek quotations without delay. The Editorial Board agreed to prepare a schedule of requirements but did not think that new printers could be brought in before January 1964. By June 1962 the Secretary and the Treasurer had met a representative of the Cambridge University Press and "no resentment at the action the Society had taken was detected". "The Press offered to continue with existing arrangements if the Society's plans to find alternative printers and publishers fell through". The Committee decided that January 1964 would be convenient for the start of a new arrangement.

The Editorial Board (with W. V. Thorpe as Chairman) had considered estimates from printing firms but the saving by then regarded as likely was not as great as they had been led to expect. The Committee at its November 1962 meeting were told that the Editorial Board doubted the calculations of financial advantage and opposed the proposed change. The Treasurer was however strongly in favour of the proposed change; the Society was overspending and a relatively small saving on the *Journal* would help to wipe out the deficit. Finally, after much debate eight members of the Committee were in favour of the change and six against. After the vote had been taken the Chairman (J. N. Davidson) aligned himself with the eight. The Committee nevertheless unanimously passed a vote of confidence in the Editorial Board, but by December 1962 the dispute between the Committee and the Board had become a crisis. The Board, meeting again in November, had "heard with interest statements by the Officers in favour of the change of printers but remained unconvinced about the wisdom of the change". The Board no longer considered themselves sufficiently independent to conduct the *Biochemical Journal* and decided to resign *en bloc* with effect from 1 January 1963. To avoid a breakdown the Board, however, were willing to continue their duties in an acting capacity if the Committee so wished, until such time as a newly constituted Board was able to take over. The length of the interim period should be as short as possible and should not extend beyond 1 September 1963.

On 11 December 1962, H. J. Rogers as Deputy Chairman presented the case of the Editorial Board and then the Chairman of Committee (Davidson) read extracts from the Rules to show that ultimate responsibility for the management of the Society's affairs lay with the elected members of the Committee. After strenuous discussion it was decided that four representatives of the Committee and four of the Board should meet on 14 December to seek a way out of the impasse. This group, under the Chairmanship of N. F. Maclagan, reached the unanimous decision to delay changes in the arrangements for one year. The Cambridge University Press, the Chemical Society and all interested parties were informed and the members of the Editorial Board withdrew their resignations. An advisory publications committee was recommended as a co-ordinating body.

The Committee met in February 1963 and duly set up an Advisory Committee for Publications consisting of the Chairman of Committee, the Chairman of the Editorial Board of the *Biochemical Journal*, the Honorary Treasurer and one Honorary Secretary, all *ex-officio* and entitled to appoint a deputy, together with three members from the Editorial Board and one from the Committee of the Society. The new body should "consider and make recommendations to the Committee of the Society on all matters concerning the publications of the Society with the exception of *Clinical Science*". It was decided that the meetings of the Advisory Committee should be formal, that minutes should be kept but that copies of them should not be generally available. It was agreed that the Honorary Secretary of the Society serving on the Advisory Committee should be its secretary and that the Chairman of the Advisory Committee should be elected annually by the Committee itself, but not from among its *ex-officio* members.

The Minutes of the Society's Committee dated 20 September 1963 recorded a decision reached unanimously by the Advisory Committee for Publications that as from 1 January 1965 the *Biochemical Journal* should be printed by the Spottiswoode, Ballantyne Company Ltd., and published by the Society using the Chemical Society as its agents. It was estimated that on the figures for 1963 such an arrangement would have effected a saving of about £3,600. The findings proved acceptable to the Editorial Board and to the Society's Committee. In January 1964 the agreements with the Chemical Society and with the printers had been signed.* Thus ended a traumatic experience in the history of the Society.

The Report of the Secretaries to the Society for 1962–1963 contains extremely discreet references to what had been perhaps the sharpest controversy in the recent history of the Society.

> "Faced with the continually rising costs of publishing the *Biochemical Journal* the Committee is continuing to seek new ways of increasing efficiency while maintaining the high standard of publication. As reported by the Honorary Secretaries at the last Annual General Meeting, the Society received an offer from the Chemical Society to assist in the sales and distribution of the *Journal*. The Committee felt that such an arrangement might be to the mutual advantage of the two Societies. Accordingly, much effort has been devoted by the Officers, Committee and Editorial Board to studying the full implications of such a change from the existing arrangements with the Cambridge University Press. Unfortunately, the latter were not prepared to continue to print the *Journal* unless they also published it. The investigations therefore involved alternative printing arrangements and it was not possible, in the time available, to make satisfactory arrangements for printing the *Journal* so that the new plan could start in 1964. The Committee has, however, set up an Advisory Committee for Publications, consisting of representatives from both the Editorial Board and the Committee, which will advise these bodies on all matters concerning the publications of the Society with the exception of *Clinical Science*."

H. J. Rogers succeeded W. V. Thorpe as Chairman of the Editorial Board at the end of 1962. The Committee placed on record the members' deep appreciation of the retiring Chairman's services.

As the *Journal* became larger the amount of work falling on members of the Board increased steadily and additional members were appointed from time to time. The Board is now a very large one but this is a result of the range of the topics discussed in papers and the number of papers.

---

* It should be mentioned that, after the distribution of the *Biochemical Journal* had been carried out by the Chemical Society for four years, the function was eventually passed to the Biochemical Society's own offices which had in the meanwhile taken lessons from their seniors.

OTHER PUBLICATIONS

Clinical Science

In 1954 the Association of Clinical Biochemists planned a meeting to be held in Edinburgh on 10 April, the day following a meeting of the Biochemical Society. It was agreed to give publicity to a joint programme and to arrange for members of either body to attend both meetings. This gave formal expression to important shared interests.

In 1956 two representatives of the Biochemical Society (E. J. King and C. E. Dalgleish) met representatives of the Association of Clinical Biochemists (N. F. Maclagan, C. P. Stewart, I. MacIntyre and A. L. Tárnoky) to consider establishing a journal of clinical biochemistry. It was agreed that the project of producing such a journal as a joint venture was sound and thoroughly desirable. Further, it was agreed that the two sponsoring Societies should be represented equally on an editorial board. The Committee of the Biochemical Society accepted the recommendations and set in train amendments to the Rules to cover setting up a new journal. A special committee made up of five members from the Association and five members from the Society was empowered to consider the detailed arrangements. Almost a year later the Committee of the Society decided that a memorandum on the project drafted by one of the Honorary Secretaries (C. E. Dalgleish), and accompanied by a questionnaire, should be circulated to all the members, then numbering about 2500. The number returned (758) was disappointing but 81 per cent of the replies were in favour and 19 per cent against the project of instituting a journal of clinical science as a joint venture (January 1958).

By this time another Body, the Medical Research Society, which already produced a journal entitled *Clinical Science*, intimated to the Officers of the Biochemical Society that its members were strongly in favour of extending the scope of *Clinical Science* in collaboration with the Biochemical Society. Informal consultations with the Association of Clinical Biochemists elicited willingness to be party to wider negotiations than those begun two years earlier.

In October 1958 a negotiating committee reached the following conclusions:

1. that *Clinical Science* should continue to be a medium for the publication of papers primarily on diseases in man,
2. that papers on pure methodology would not in general be accepted,
3. that there should be parity of editorship between the Medical Research Society and the Biochemical Society,

4. that the Committee of the Medical Research Society should recommend the Trustees of *Clinical Science* (then Sir Francis Fraser and Sir Alan Drury) to increase their number to four, two of whom should be representatives of the Biochemical Society, and
5. the Trustees would be the legal owners but would have no concern with the day to day running of the journal.

The structure of the Editorial Board and the duties of a managerial committee were set out and arrangements were made for drawing up legal agreements. It was decided also that the Association of Clinical Biochemists should participate through the Biochemical Society in the work of the Editorial Board, but not in ownership of the journal. Later in 1959 draft agreements were drawn up and ratified in 1960. Sir Charles Dodds and Sir Charles Harington agreed to be the Biochemical Society's two Trustees.

The Society's representatives on the Editorial Board were N. F. Maclagan, C. H. Gray, N. H. Martin and F. T. G. Prunty. Each Society transferred £1000 to a joint account for financing *Clinical Science*. The Committee of Management agreed that the Cambridge University Press should publish the journal and Messrs Shaw and Sons should print it.

In 1963 it was reported that in its first year under joint auspices *Clinical Science* had lost £200, in its second year £251 and in its third year £1374. It was expected that the loss in 1964 would be a more 'normal' one of £500.

In 1963 the Committee of the Biochemical Society received a report from the Committee of Management of *Clinical Science* recommending that as from 1 January 1965 the Academic Press should print *Clinical Science* and Blackwells Scientific Publications Ltd of Oxford should publish it. Both the Biochemical Society and the Medical Research Society agreed. It was further decided that the price to non-member subscribers should be increased from £4 to £5 per annum. The price to members of either Society was increased from £2. 10. 0d. to £3. 10. 0d. per annum. By 1968 the subscription rate was £5 per volume of three parts or £10. 0. 0d. per annum.

At the time of writing the Editorial Board consists of seven members from the Biochemical Society (including D. N. Baron as Chairman) and seven members from the Medical Research Society (including J. M. Ledingham as Deputy Chairman).

Clinical Science has now established itself firmly in a highly competitive field in which many new journals have been founded. The association with Blackwells Scientific Publications Ltd, Oxford, has proved very successful.

Two-thirds of the sales of 1868 copies (April 1968) were through the trade and 626 to members (89 Biochemical Society and 537 Medical

Research Society). The geographical distribution of sales was: United Kingdom 42 per cent, U.S.A. 24 per cent, Europe 17 per cent, Commonwealth 17 per cent, others 7 per cent.

During the past four years about 100 papers a year averaging 10 pages in length have appeared, but the rejection rate has remained high and in 1967 reached 77. In its report for the year 1967 the Editorial Board reported that the problem of medical ethics had caused it much concern; "a number of papers have been rejected because it was obvious that the work reported in no way satisfied the criteria recommended by the Medical Research Council for investigations on human subjects".

Despite the difficulties, and the rejection rate which must have increased the labours of the Editorial Board, the average delay in publication was 9.0 months in 1967.

Clinical Science, owned and administered jointly by the two Societies, has now four Trustees, a Committee of Management of 10 members and an Editorial Board of 14 members. Less than two-thirds of the papers published in 1967 originated in the United Kingdom. The fact that over the past four or five years the number of papers submitted has been reasonably constant is not an indication that the output of papers in clinical science is levelling off. Rather it is due to the fact that new journals are competing for papers. The Editorial Board is however satisfied that the quality of the material submitted is being maintained.

Essays in Biochemistry

This annual publication was intended to be "read with pleasure and profit by senior students and lecturers in biochemistry. Each essay presents an overall view of one aspect of the subject, indicating its origin, present status and likely future development".

The decision to undertake the project was taken by the Committee in July 1963 and in September P. N. Campbell and G. D. Greville were nominated by the Advisory Committee for Publications to act as editors.

In October that year an eminent member of the Society who had been invited to contribute an essay pointed out that "in her opinion under Rule 13 the Committee was not empowered to publish such a series until its establishment had been approved at a General Meeting of the Society". The Advisory Committee for Publications did not regard the intended paperback series as coming within the meaning of the word 'journal' under Rule 13. There were further exchanges by correspondence and eventually the matter was discussed at the Annual General Meeting in the Spring of 1964. The proposal was carried and an agreement with the

Academic Press was signed. The discussion had however served a useful purpose and a clarification of the scheme was approved by the Advisory Committee for Publications. The *Essays* were intended to be "based on" *Physiological Reviews* but shorter, since "over-documentation tended to make reading difficult, references would be limited but adequate".

Four volumes of *Essays in Biochemistry* have now appeared. They contain 19 essays averaging 40 pages in length the total cost being 96s. or 76s. 6d. at the special price to members for single copies. Press comment on the series has been very encouraging and the success of the venture seems assured.

Occasional Publication

An interesting innovation was the publication in July 1969 of a booklet called *Biochemistry, 'Molecular Biology' and Biological Sciences* which was the report of a subcommittee (Sir Hans Krebs, Chairman, W. N. Aldridge, K. S. Dodgson, S. R. Elsden, G. A. D. Haslewood, A. P. Mathias, D. C. Phillips, R. M. S. Smellie) set up by the main Committee to consider the Kendrew Report on molecular biology (H.M.S.O., Cmd. 3675). The Krebs Report, as it might be called, thoughtfully reviewed the progress and standing of biochemistry in schools, universities, research institutes and industry and made suggestions for its future development.

MEETINGS

ORDINARY MEETINGS

From the earliest days of the Society onwards the main activities of the Society were holding meetings and publishing the *Journal*. From 1913 to 1921 attendances at meetings averaged 40–50 and consisted mainly of people residing in the London area. Even a good deal later N. W. Pirie has mentioned that Cambridge biochemists attended only the Cambridge meetings; "we did not even go to Oxford". There seems to have been a certain aloofness but it must be remembered that between 1919 and 1946 salaries in British universities were low and travelling expenses to attend meetings were rarely granted. It was only after 1947 that provision was made by the universities to refund expenses "incurred in reading a paper" to a meeting of a learned society and even then the total was strictly rationed. There has been much improvement since and the societies themselves have become somewhat more generous in this respect to members of committees.

No doubt other factors have played a part—notably the rapid growth of biochemistry and in the number of practising biochemists—in the greatly increased attendances at meetings. Indeed some meetings have become very large indeed.

An incident of some importance took place in 1926 when Harington protested to the Committee against the unauthorized publication in daily newspapers of his paper read at the March meeting. In June the Committee decided that in future it should be announced at each meeting of the Society that the proceedings were to be considered private and should not be reported to the Press without the consent of the authors. This principle has been adhered to ever since and on the whole press reports about biochemical research have not been irresponsible or sensational.

In 1930 the Committee tackled the thorny problem raised by speakers at meetings who exceeded their time allowance. There was a good deal of discussion; Harden and Gardner wanted to leave it to the Chairman to apply the 15 minute rule strictly. Chibnall was anxious to maintain the informal nature of the Society's meetings. But it was decided by five votes to three to use an electrical device which showed a red light at the end of 12 minutes and at the end of 15 minutes actuated a buzzer. The Minutes leave it to be concluded that the decision was put into effect with good results.

Soon after the war the Committee arranged for abstracts of communications to be pre-circulated; it also passed that speakers were to be allowed

10 minutes to deliver their presentation with five minutes allotted for discussion, this to be lengthened at the discretion of the Chairman, but the Meetings Secretary was to be responsible for seeing that the allotted five minutes was not exceeded. It seemed that the discretion of the Chairman was considerably curtailed! At this period (1947 onwards) the local organizers of meetings felt it was incumbent on them to display ingenuity in devising combinations of flashing lights, buzzers and bells to deter speakers from going on after their time was up. The Committee also exhorted members to see that lantern slides and diagrams for projection by the epidiascope could be seen clearly by the audience. In this connexion it is right to recall that A. L. Bacharach, who rendered great service to the Biochemical Society, the Nutrition Society and other scientific bodies, was more than anyone else responsible for a notable improvement in the slides shown at meetings. What he said was obvious enough but it needed saying and needs to be repeated periodically! As a referee and as a member of the Editorial Board of the Nutrition Society he could pounce upon slipshod writing with such zest that he was always forgiven. At meetings of the Biochemical Society, which he attended with great regularity, there were none to match him in opening a discussion with a pertinent comment or question.

There were of course occasions when discussion fell flat. Pirie remembers a Cambridge meeting on 25 October 1924 when J. B. S. Haldane, J. Hicks and C. Watchorn discussed the effects of inducing acidosis in Haldane—always ready to act as a test organism. Hopkins as Chairman was complimentary and invited discussion, but there was none. "Even at that time people were rather frightened of Haldane." Then Sir Charles Martin rose and said "These are very interesting and important experiments, so interesting and important indeed that they ought to be repeated on a normal subject". Haldane attended many meetings and was liable from the back row of the lecture theatre to ask questions. It was not always easy to decide whether the speaker had been floored or had not heard the question. Be that as it may it was a sign of the times when in 1948 Haldane complained formally and officially to the Committee about overcrowding at a Symposium.

Increasing attendances prompted members to complain that names could not be identified with faces and the Committee seriously considered a suggestion that badges bearing names should be worn at meetings. The idea was rejected in 1950, but the problem of identification at meetings was again discussed in 1958. It was agreed that labels were desirable but a proposal to use them was again condemned as 'impracticable'. But a year later it was agreed that "labels should be provided at all meetings on a trial basis for one year". In 1960 it was decided that the use of such labels should be made permanent, the trial period having proved their usefulness.

In 1960 the Committee considered a letter complaining about the poor quality of lantern slides shown at a large meeting of the Society held at Cambridge. The Secretary stated that he had written "to offending members and had received appreciative letters". Charmed (or disarmed) the Committee decided that this "was likely to be the most effective method of dealing with the matter", although a few Heads of Departments may have had their interest revived.

In 1961 a new issue arose about reporting some of the Society's meetings. When an important discussion took place it was natural for the Press and particularly the scientific weekly press to want to comment on anything newsworthy. The Committee decided to remind all members that the meetings were private and were not to be reported to the Press without the permission of the authors and of the Secretaries. If any member were asked to draft a report he should first obtain permission from the Meetings Secretary. A letter on those lines was sent to *Nature* but the publishers rather promptly pointed out that they could in no way countenance even the appearance of such a curtailment of the freedom of the Press. This was accepted by the Committee of the Society and it was agreed that "there would be no delay in scrutinizing draft reports of appropriate meetings". The Society is on strong ground in laying down that Press reports on individual papers should have the permission of the authors but the requirement that the Secretaries should intervene is less easy to defend.

By the 'sixties' the organization of meetings had become quite complicated and much work was thrown on local organizers. Discussions and colloquia had become regular features. The number of papers submitted in 1963 rose to 326 or 36.2 per meeting. Such heavy programmes reflected the rise in membership and of course the growth of biochemical research in the country. It became necessary to use two or more lecture theatres simultaneously. This had the disadvantage that although the total attendance at a meeting might be very high, the actual audience for many papers could be meagre, depressing alike to the speakers and audience. Many members of the Committee felt that the tradition that any member should be allowed to present his work in 10 minutes and to have an abstract printed was too valuable to be readily given up. It was, however, admitted that few members of an average audience could "make the attempt to understand a series of communications on a bewildering variety of topics". This was thought to explain the fall in lecture theatre attendances at ordinary meetings. Nevertheless the opportunity to present a piece of research was an essential factor in membership.

The dangers of the situation were sharply illustrated at the Oxford meeting in September 1963; 700 members attended this meeting at which

111 papers were read in five 'parallel' sessions, but actual audiences were sometimes very thin. In addition there were colloquia and a discussion symposium. Apparently the very high participation put some strain on the colleges and the Committee, following complaints, had a somewhat acid Minute on spartan accommodation. The Secretaries were empowered to write to Heads of Departments at Oxford and Cambridge and not surprisingly they were advised to contact the Domestic Bursars' Committee.

It will be seen that a meeting of the Society has ceased to be a simple and informal gathering of a small group of like-minded biochemists. Each meeting is to some extent a conference and considerable demands are made on the local members as well as on the Headquarters and Officers of the Society. There is still a substantial majority in favour of holding meetings at different centres all over the British Isles and the number of such centres has increased quickly with the creation of new universities and new research institutes. The organization of each meeting needs a good deal of planning well in advance and a big meeting with over 100 papers to be read makes heavy demands in many ways.

The successful inauguration of Groups and the organization of collaboration with other societies, to say nothing of the growth of the Federation of European Biochemical Societies, means that lines of communications are singularly unimpeded. Few will deny that all these activities are justifiable but the individual scientist may soon find that his personal participation may perforce be subject to strict rationing in the interest of his other commitments.

DISCUSSION MEETINGS AND SYMPOSIA
(see also page 23)

H. J. Channon in December 1934, addressing the Committee, spoke in favour of Discussions and moved a motion:

"That consideration be given to the question of the Society holding meetings for the discussion of specific biochemical problems in place of, or in addition to, the ordinary meetings as at present held."

It was pointed out that the Chemical Society had arranged a number of discussion meetings and the Secretary was asked to enquire unofficially:

1. Whether members of the Biochemical Society who were not fellows of the Chemical Society were entitled to attend the discussions of the Society on subjects of biochemical interest.
2. If in the affirmative, whether there would be any objection to us notifying our members of the event.
3. Whether it would be possible for members who were not fellows of the Chemical Society to purchase reprints of the discussions.

Three months later it was reported that the "Biochemical Society would be officially invited to help arrange and take part in biochemical discussions which the Chemical Society would hold", and later that year a topic was proposed, namely, 'The chemistry and biochemistry of lipoids'.

In 1937 the Biochemical Society decided to hold one meeting a year for discussions on subjects of general interest, such a meeting to be additional to the normal eight meetings of the Society. In 1940 a joint discussion of the Faraday, Physiological, Biochemical and Chemical Societies took place on 'Chemical structure in relation to membrane permeability'.

In November 1941 there was a discussion meeting of the Society at the Courtauld Institute on 'The mode of action of chemotherapeutic agents'.

In 1942 the Committee recorded the "apparent widespread desire of members of the Society for discussion meetings". Two topics were suggested 'Metal-containing pigments of biological importance' and 'The theory and application of chromatographic methods to biological problems.' The principle was accepted that a summary (not exceeding 700 words) of the opening papers should be pre-circulated to members of the Society. A discussion on 'Tetra-pyrrolic pigments' was held in 1943 and there were many requests for it to be published as a pamphlet. One very influential member of the Committee advised against this on the following grounds:

"1. The present shortage of paper.
2. Pamphlets rapidly disappear from circulation thereby losing any value they possess.
3. Publication in any case would create a precedent which may not be desirable and to which contributors to future discussions might object."

In May 1944 a special subcommittee recommended that discussion meetings should be a definite part of the meetings programme and that at least two discussion meetings should be held each year. It was also decided that for the time being such meetings should be held in London, Oxford or Cambridge. Non-members were to be invited to contribute and it was suggested that the text of contributions "should be issued before-hand to those who apply". The Committee rejected the recommendation to limit discussions to the places named and the exclusion of Scotland seemed to have been a sore point. N. W. Pirie, who had been asked to sound out possible speakers for several suggested topics, did not find his task at all easy, presumably because war-time commitments were a heavy burden to potential contributors.

Nevertheless, by June 1945 discussion meetings were a going concern and were already attracting favourable comment.

DISCUSSION MEETINGS 1941–49

Date and Chairman	Title and Some Speakers
November 1941 (E. C. Dodds)	Mode of action of chemotherapeutic agents. (G. M. Findley, Fleming, Warrington Yorke, F. Hawkins and Dale)
November 1942 (I. M. Heilbron)	The theory and applications of chromatographic problems to biological problems.
November 1943 (J. Barcroft)	The tetra-pyrrolic pigments. (D. Keilin)
December 1944 (C. R. Harington)	Joint meeting with the Pathological Society and the Royal Society of Medicine on cancer. (Dodds, Haddon, G. M. Bonsor and others)
April 1945 (R. A. Peters, L. J. Harris)	The vitamin B complex.
November 1945 (R. A. Peters)	The chemical basis of cell structure and function.* (Fabre, Fromageot, Wurmser, Jorpes)
February 1946 (E. L. Hirst, A. R. Todd)	Amino-sugars and uronic acids in nature. (G. Blix, E. G. V. Percival)
November 1946 (P. Fildes, A. L. Bacharach)	Joint meeting with Society for General Microbiology Quantitative biochemical analysis by microbiological response.
February 1947 (A. C. Chibnall, C. Rimington)	The relation of optical form to biological activity in the amino-acid series.
December 1947 (M. Dixon, R. A. Peters)	The biochemical reactions of chemical warfare agents.
October 1948 (E. Baldwin, R. L. M. Synge)	Partition chromatography and its application to biochemical problems.
February 1949 (J. B. S. Haldane, A. Neuberger)	Biochemical aspects of genetics.

*This meeting was one of the early occasions with an international flavour. Professor Fabre on behalf of the French Biochemical Society presented a Pasteur Medal with the inscription "La Société de Chimie Biologique à la Biochemical Society, Témoinage d'Amitié 10.xi.45".

Professor Peters accepted this token of friendship with thanks to Professor Fabre and his colleagues for their kindness. (See *Biochem. J.*, 1946, **40**, 1).

Members were provided with abstracts (or synopses) of the contributions but there was widespread feeling that a fuller account was needed. At the time paper was rationed and the shortage inhibited action. A year later R. T. Williams, who had been appointed organizer of discussion meetings, made a statement to the Committee in favour of the proposal that "complete reports of the Society's Discussion meetings be printed". E. J. King, supporting, said that the *Journal* did not possess the necessary supplies of paper and urged that the Reports should not form part of the *Journal*. It was agreed to publish the Reports as separate volumes. The Cambridge University Press quoted £105 for 100 pages or 4s. per copy if 500 were printed or 2s. 6d. per copy if 1000 were printed! Sales were at first slow and there was anxiety about the venture but eventually Symposium I was sold out (1000 copies) and made a profit. The agreed policy was to print the contributed papers but not the actual discussions and the series was entitled *Biochemical Symposia*. There were to be no free reprints but each contributor was to have one free copy and permission to buy up to 100 reprints. By the time the fourth Symposium was published the outlook for sales justified printing 2000. The Symposia were sometimes reported in other journals as unsigned summaries and some authors complained that such articles were inadequate and not always accurate. In recent years, however, such summaries have been very competently prepared.

R. T. Williams remained Symposium Organizer until 1955 and by that time the series was fully established as an important part of the Society's activities. The task of organizing each Symposium had been arduous and the subsequent production of the printed version involved much work. The symposium subcommittee decided in 1954 that two symposia should be held each year and that overseas speakers could be invited. They recommended that E. M. Crook be appointed Symposium Organizer from April 1955. A further recommendation limited tenure of the office to seven years.

In 1955 the Society collaborated with the Society of Chemical Industry in arranging a symposium on 'Hypotensive drugs'. In the following year the Society was represented on a committee of the British Pharmacological Society organizing a symposium on 5-hydroxytryptamine. In 1958 J. K. Grant succeeded Crook as Symposium Organizer and the symposium subcommittee was reconstituted with new terms of reference, namely, "to consider and make recommendations to the Committee on all matters concerning subjects, times and places of symposia and publications arising therefrom". It was agreed that although hard and fast rules would be inappropriate, not more than a quarter of the speakers should be from laboratories outside the British Isles.

It is only necessary to examine the titles in the succession of symposia to appreciate that the series has become a notable one, and indeed memorable papers have been submitted by speakers from many countries. From 1960 onwards the Society was represented by H. McIlwain on a Co-ordinating Committee for Symposia on Drug Action. This work was very successful. In 1963 J. K. Grant was replaced by T. W. Goodwin as Symposium Organizer.

In 1967 a symposium to honour Sir Hans Krebs on his retirement was arranged at Oxford under the title 'The metabolic rôle of citrate'. Messrs Boehringer offered a financial contribution towards this.

It has been the Society's policy not to lose money on symposia but nevertheless to keep prices down and the policy of offering them for sale in paper covers or hard-board has encouraged the sale both to individuals and to laboratories.

T. W. Goodwin will by the end of 1970 have completed his seven years as Symposium Organizer and R. M. S. Smellie will succeed him. They will work together in 1969.

Besides formal symposia, nine colloquia were held in conjunction with Ordinary Meetings in the Session 1967-68. These covered a very wide range of topics. The informal discussion meetings started in 1966 were continued in 1967-68 and three were held during the year on afternoons or evenings preceding Ordinary Meetings:

Functions of lysosomes, Cardiff, January 1967
Metabolic rôle of vitamin A, Dublin, March 1967
Mechanism of action of steroids, St Andrews, June 1967.

These informal discussions proved enjoyable and undoubtedly served a useful purpose.

SYMPOSIA TITLES

No. 1. The Relation of Optical Form to Biological Activity in the Amino-Acid Series.
No. 2. The Biochemical Reactions of Chemical Warfare Agents.
No. 3. Partition Chromatography.
No. 4. Biochemical Aspects of Genetics.
No. 5. Biological Oxidation of Aromatic Rings.
No. 6. The Biochemistry of Fish.
No. 7. The Biochemistry of Fertilization and the Gametes.
No. 8. Metabolism and Function in the Nervous Tissue.
No. 9. Lipid Metabolism.
No. 10. Immunochemistry.
No. 11. Biological Transformations of Starch and Cellulose.
No. 12. The Chemical Pathology of Animal Pigments.

No. 13. The Biochemistry of Vitamin B_{12}.
No. 14. Nucleic Acids and Protein Synthesis.
No. 15. Metals and Enzyme Activity.
No. 16. The Structure and Function of Subcellular Components.
No. 17. Glutathione.
No. 18. The Biosynthesis and Secretion of Adrenocortical Steroids.
No. 19. Steric Aspects of the Chemistry and Biochemistry of Natural Products.
No. 20. Biochemistry of Mucopolysaccharides of Connective Tissues.
No. 21. The Structure and Biosynthesis of Macromolecules.
No. 22. The Structure and Function of the Membranes and Surfaces of Cells.
No. 23. Methods of Separation of Subcellular Structural Components.
No. 24. The Control of Lipid Metabolism.
No. 25. Aspects of Insect Biochemistry.
No. 26. Instrumentation in Biochemistry.
No. 27. Metabolic Rôles of Citrate. (Organized as a tribute to Sir Hans Krebs.)
No. 28. Porphyrins and Related Compounds.
No. 29. Natural Substances formed Biologically from Mevalonic Acid.

SYMPOSIA ON EDUCATION

The Biochemical Society has not confined its activities to publishing original papers and holding meetings devoted exclusively to presenting new work. It has also held meetings on topics like the training of biochemists and two monographs with that title have appeared.

A colloquium was held at Oxford on 13 July 1961 with K. S. Dodgson in the Chair, and a full account (43 pages) of the proceedings came out later that year and was sent to all members. A second colloquium on the training of biochemists was held at Aberystwyth on 15 September 1966 and also published as a monograph (45 pages). There was wide agreement that the colloquia and the monographs provided valuable information and that the exchange of views had been worthwhile. It was not suggested that the Committee endorsed all that was said, indeed the Society was more concerned to encourage informed discussion than with advocacy of particular causes. But as Dodgson said in his Chairman's remarks (1961):

"During recent years the Biochemical Society has taken an increasingly wide view of its responsibilities for the well-being of the subject . . . a special difficulty of biochemistry is that its potentialities as a career are not fully appreciated in the schools although it is realized that the shortage [of biochemists] represents only a part of a much larger problem extending throughout the whole of science. . . . However, recent correspondence with colleagues overseas has convinced me that no one is yet certain as to the most appropriate way of training biochemists."

S. V. Perry considered biochemistry as an Honours Degree subject and reviewed the history of courses in Britain. In a closely reasoned presentation he said that "a balanced cover of modern biochemistry makes great demands on staff. . . . All teachers of biochemistry must at times have felt oppressed by the tremendous expansion of knowledge of the subject and the difficulties of ensuring that Honours courses in biochemistry keep up with the latest advances. It is in fact essential for the effort to be made as it often leads to a simplification and helps to make sense of the immense mass of material which accumulates." . . . "Clearly this process cannot go on indefinitely and we must set our sights on some objective in an Honours course." Perry was a little anxious that the Honours course should not "suffer as an exercise in pure intellect" because of its large descriptive content.

B. Spencer compared the British system with that commonly in use in the United States of America and concluded that "the strength of the American system, as it is at present, is the opportunity for the student to obtain a fuller scientific background than his British counterpart, before being plunged into the complexities of advanced biochemistry" [as a research worker.] The two papers were followed by an interesting discussion.

W. F. J. Cuthbertson presented a paper on the graduate biochemist in industry and argued that although the basic training requirements were well met by the university schools of biochemistry, more specialized industrial needs might be met by placing special emphasis on particular aspects of the subject at different places. Too few biochemical graduates were taking up industrial appointments and too many were retained in academic posts. G. A. Snow said that industrial concerns were looking for biochemists of the highest calibre and preferred to take them after they had gained a Ph.D. degree.

Margaret Kerly gave a factual account of university and departmental entrance requirements in 1961. Since then a number of changes have occurred.

D. D. Woods, speaking of the teaching of biochemistry at Oxford, recalled that the first batch of students taking biochemistry as a separate Honours degree subject graduated in 1952. The system was later changed... "our objective is to produce not one type of biochemist but as broad a spectrum as possible". The Oxford student had the benefit of the tutorial system but the University course aimed at a broad outlook.

R. A. Morton mentioned a Royal Society Report by an *ad hoc* Biological Research Committee presented to the Advisory Council on Scientific Policy and drew attention to the section on the teaching of biology. . . . "Recent advances were cutting right across the old lines of separation between the different subjects. . . . Honours courses in biological subjects might in the future require a greater variety of ancillary subjects—more numerous but perhaps shorter 'supporting' courses".

E. M. Crook discussed the M.Sc. course in Biochemistry and outlined developments in advanced lecture courses which have earned recognition and support from the Science Research Council. It was already accepted in many university departments that the first year of a three-year postgraduate course might well contain a substantial number of advanced lectures.

F. C. Happold, concluding the discussion, returned to educational fundamentals when he said that the *education* of biochemists embraced their training but was more difficult to achieve. He ended with an amusing and pointed Yorkshire anecdote which will reward those who re-read the monograph.

The Society held an evening discussion at the Middlesex Hospital Medical School on 16 April 1964 on the place of biochemistry in the new universities. An excellent report (by E. A. Dawes) appeared in *Nature* (1964, **203**, 4941). Sir Rudolph Peters in his opening remarks as Chairman pointed out that "nowadays many aspects of biochemistry masqueraded under other titles. The new term 'molecular biology' had been introduced by Astbury and was perhaps valuable for getting money, but the claim that biochemistry only dealt with small molecules was just not true—we had always dealt with substances like proteins".

R. A. Morton surveyed the growth of biochemistry as an academic discipline. The interpenetration of the biological sciences called for psychological readjustment and reconstruction of teaching, organization and administration on a scale easier to realize in a new university than in an established one. "There is, however, a hard core of biochemistry which cannot safely be watered down".

J. N. Davidson said that biochemistry had tended to suffer from being a hybrid subject. He welcomed the growing interest of chemists and biologists in biochemistry but was utterly opposed to the trend that biochemists might develop into mere laboratory assistants helping "big brother biologist" to solve his problems. E. A. Dawes reported on what was happening to biochemistry in several new British universities and noted how closely the suggestions of the Royal Society Report (*Annual Report Advisory Council on Scientific Policy* **1961–62**. London, HMSO 1963) were being followed. Multi-professorial departments housing under one roof a variety of biological disciplines were being set up. Some speakers, however, feared that the basic biochemistry course might become too flimsy.

H. L. Kornberg spoke of developments at Leicester with flexibility in courses covering various aspects of cell biology as the main aim. A. J. Birch discussed the development of biological chemistry at Manchester with a new course leading to an Honours degree on a broad base.

T. A. Bennet-Clark dealt with the rôle of biochemistry within a school of biology and discussed the scheme in operation at East Anglia. Biochemistry was fundamental and must be incorporated in the courses, but "this could only be achieved by elimination of dead wood from the more classical aspects of the biological parts of the curriculum".

The meeting was universally adjudged to have been a timely and instructive one and the large attendance reflected the widespread interest in developments in the new universities.

A little later a new Committee returned to the subject of the training of the biochemist and G. R. Barker was asked to arrange another colloquium. Accepting the over-riding need for the process to be educational, Barker submitted that training could be "a means to a variety of ends" related to the state of scientific knowledge and of technology, but he discerned a wish to see "both a chemical quart and a biological quart in a pint pot". Some of the replies to his enquiries stipulated that "postdoctoral experience should be relevant to the firm's work" indicating perhaps "a desire to buy ready-made ideas instead of putting faith in a man. . . ."

J. T. Edsall dealt with undergraduate training in relation to biochemistry. He recalled that at Harvard in 1910 L. J. Henderson started giving a course for undergraduates in biochemistry with laboratory work, and that he himself took the course in 1923 and followed it up by taking F. G. Hopkins's Part II course in biochemistry during its first year (1924–25). Edsall went on to outline the explosive development of the science in the past 40 years; he stressed the present-day importance of chemistry as well as general biology and noted a danger "of superficiality in an undergraduate programme in biochemistry". He went on to say "Such a biochemistry major is obviously an exacting programme" and "the level would depend on the quality of previous education" at school. Undergraduate students of biochemistry might attend "a mere collection of diverse courses with little unity or coherence". Edsall quoted with obvious anxiety from a *Nature* editorial (1966, **211**, 554) "British universities remain curiously inflexible in what they ask of intending entrants to chemistry departments. . . . One immediate absurdity is that only a tiny handful of British graduates in chemistry have ever studied biology except at an elementary level".

A. P. Mathias discussed the functions of M.Sc. courses in biochemistry under the headings

Conversion courses	for those moving from another discipline into biochemistry
Extension courses	to augment studies in biochemistry already begun
Specialization courses	which concentrate on a narrow aspect of the subject.

Considerable developments had taken place in this field since 1961. The lecture courses leading to a Master's degree have come to stay because they are educationally sound and meet real needs.

R. A. Morton surveyed the training of the postgraduate student. He said that "every research student is an important individual selected on the basis of high promise and the relationship between him and his supervisor is a unique one". The initial investigation should be chosen carefully because reasonable progress could generate self-confidence. The major problem would involve reconnaissance, periodical pauses to take stock, and the research student would reach his own solution of the problem of coping with the literature. The supervisor's main task was to foster the development of critical judgment without "over-extending the pupil-to-teacher relationship". The postgraduate student needed to strike a reasonable balance between having a sophisticated approach to instruments and an appreciation of the shifting equilibrium between scientists and their technician colleagues. Training should not neglect safety precautions, and the cost of research as well as the nature of waste should be understood. Finally the educative value of studying official reports relevant to biochemists and nutritionists was illustrated by examples.

Postgraduate work now looms larger than ever and it is a sign of the times that The Royal Society has recently (1968–69) published Reports on Chemistry, Physics, Biology, and Engineering and Technology prepared by subcommittees of a main committee appointed to examine postgraduate training in Science and Technology.

C. H. Selby discussed the approach to biochemistry from school science. He posed the question: "How can we widen the scope of teaching in the sixth form without increasing the burden of detailed subject study and at the same time allow the important development of a wide range of non-examination subjects to continue." He mentioned J. E. Spice's proposals for an amalgamation of Chemistry and Physics and quoted from J. W. S. Pringle on "the two biologies". "The first biology considers how organisms came to be what they are and what function their parts have in relation to the whole, the second biology analyses what existing organisms are and how what they and their parts do may be described in terms of physics and chemistry". Selby concluded that "any chemistry syllabus developed to support biological studies . . . . could not fail to be of considerable interest".

J. B. Jepson had organized a display of teaching aids appropriate to biochemical education and the interesting descriptions and comments were put on record. The monograph ended with statistical appendices.

A practical demonstration of the Society's concern was the preparation of a pamphlet entitled *Careers in Biochemistry*. The prime mover in this excellent project was K. S. Dodgson and the publication, which has been freely distributed, provides much information useful particularly to sixth formers, careers masters and parents. It is the settled policy of the Society to keep the booklet as up-to-date as possible.

The Society's concern with education was further shown by the appointment of a subcommittee under the chairmanship of D. N. Baron to prepare a memorandum on the place of biochemistry in undergraduate and postgraduate medical education. The memorandum was submitted to the Royal Commission on Medical Education presided over by Lord Todd. Later D.N. Baron, J. B. Jepson and P. J. Randle represented the Society at a meeting of the Royal Commission and gave oral evidence. The text of the memorandum was published in the *Year Book and Annual Reports for 1967* together with a report on the oral submission.

The broader problems associated with the place of biochemistry in universities, new or old, are difficult and are made more complicated by the rapid growth of the subject. The rôle of undergraduates in solving such problems cannot be large but the young postgraduate student may well have an important contribution to make.

A further sign of the Society's educational involvement was the allocation by the Committee of funds for distribution to University Departmental Biochemical Societies so that the expenses of special lectures could be met. Individual grants (up to £20) amounted to 11 in 1965–66, 15 in 1966–67 and 21 in 1968–69.

Although the founders of the Biochemical Society could scarcely have imagined how much their successors were to be interested in education the commitment must be applauded. Successive Committees have decided that the Society has responsibilities to shoulder.

COLLOQUIUM ON BIOCHEMISTRY IN INDUSTRY

A very successful colloquium organized for the Society by G. A. Snow was held in Manchester on 11 November 1965. In his foreword to the published account Snow said that on 1 July 1965 the Society had 2,233 members resident within the British Isles; of these 294 could be identified as employed in industry and 30 in industrial research associations, altogether

14.5 per cent of the British membership. The numbers of non-members probably almost equalled those of members, but the total number of industrial biochemists lay between 450 and 550, distributed as follows: pharmaceuticals 50 per cent, foods 15 per cent, drinks 10 per cent, agricultural products 10 per cent, soap, tobacco, rubber, fine chemicals etc. 15 per cent.

> "Two factors have tended to increase the industrial requirement for biochemists in the last three or four years. One is the establishment in this country of new research units, especially by foreign pharmaceutical firms, and the other is the increasing demand for biochemical information on drugs, food additives and agricultural chemicals. . . . The need for biochemists in industry has not yet been fully met and . . . scarcity may continue for a year or two. . . . The new research units have taken in mainly young men and vacancies arising through promotion or retirement will be fewer than in organizations where the age spread is more even. There is little sign at present of any far-reaching change in the situation and the indications are that the number of biochemists entering industry over the next few years is unlikely to exceed the present rate of intake".

D. F. Elliott of CIBA Laboratories discussed the place of biochemistry in the pharmaceutical industry and J. D. Coombes spoke on 'Biochemists—future famine or flood'. He noted in recent years a marked trend towards biochemistry as a first degree subject. In 1963 out of 195 first degree graduates only 13 entered industry, and of 118 higher degree biochemistry graduates only eight entered industry. "It seems incumbent upon all of us in industry to ask ourselves why this should be so."

A. H. Cook dealt with 'Integration of biochemical work with wider research outlooks'. He drew a distinction between research and investigation, and under the heading 'The biochemist's outlook', called attention to the very wide scope of biochemistry and the dangers of an unbalanced attitude towards the impossible task of keeping up with the literature.

> "Biochemical training in our universities will become more valuable when it still more actively leads to a practical outlook on this question."

Discussing the sense of purpose in biochemical research, Cook thought it sometimes did not loom very large in the minds of any of those concerned in academic biochemistry. This could to some extent be remedied by the university biochemist as a consultant to industry.

> "One has indeed positively to integrate the biochemist's part if he is not in this particular context to become a kind of technical outstation for the organic chemist. If this is allowed to happen, he soon and very understandably tends to become discouraged and the whole project loses momentum."

> "Now it would be irresponsible to allege that experienced biochemists often engage in purposeless work. However, the very diffuseness and extent of biochemistry make it easy for biochemists with less experience to miss or lose that sense of purpose. Thus I was interested to hear of two pieces of research, each of three years' duration, on the uptake of individual amino-acids by different but related species of organism. My special recollection is, however, that neither of those doing that work were able to give a single word as to why they were

doing it. Perhaps those in charge of them knew, but obviously the purpose, if it were there at all, did not loom very large in the minds of any of those concerned. This, I think it fair to add, is the frequent impression, however erroneous it may be, of much so-called academic biochemistry."

Cook also spoke on 'Academic versus industrial biochemistry'.

"It is easy to understand from that kind of approach how the idea can grow up between biochemists in industry and those in non-industrial establishments that the kind of activity of each group is, in its own estimation, superior to that of its counterpart. There can be little doubt that there exists a strong current of feeling of that kind and its consequences are often unfortunate. Thus the young biochemist taking up his first post, often understands that the university post is necessarily very desirable while a comparable post in industry should only be accepted by force of circumstances. Many of us will recall instances where the potential industrial biochemist is almost a subject of commiseration in the academic department he is leaving, having to accept that not everyone can be favoured and that inevitably some must find themselves in lesser posts. Surely the remedy for this tendency is to ensure more contact between the two groups and I often think that of the various kinds of contact perhaps the most fruitful is that brought about by the academic biochemist acting as a consultant to industry. Some already do so, others would probably be prepared to and the greater initiative must of course come from industry since it will be meeting the modest financial demand. We in this country are probably much less active in this way than some European countries where the Teutonic university system often leads to the industrial biochemist being at the same time *Privatdozent* or *Ausserordentlicher Professor* in the university. In this respect we are still more behind the United States where the frequency of consultancies and the the mutual respect they engender leads to the industrial career being considerably more favourably regarded".

F. A. Robinson spoke on 'Biochemists as managers' and discussed the expression 'backroom boys'.

"These are the typical 'backroom boys', a phrase that I find particularly distasteful, although alas frequently justified. It was coined during the war as a complimentary expression, and evoked a picture of a magician-like figure performing miracles to save the country from destruction. Nowadays the phrase seems to me to be more than a little patronising. It implies a group of men deliberately shut away from the world either because they must be protected from the rigours of economic reality or because they are too ingenuous or too ignorant to be allowed to share in the management of an industrial enterprise. Unfortunately both implications may be right in many instances. . . .

"So I want in this talk to challenge any biochemists who have the back room mentality, by calling attention to the commercial facts of life and inviting them to widen their horizons. . . .

"That scientists in general and biochemists in particular do not play a more active role in management is largely, I think, their own fault. In the first place biochemical research is so very, very interesting. The exciting developments of recent years such as the unravelling of the genetic code, the isolation of new enzymes, the determination of structure and synthesis of peptides, and even proteins, the nature of the mitochondrion and so on are activites that can satisfy the most enterprising of scientists and challenge the most original of minds. Why then bother with anything else? Secondly, scientific research is nowadays so specialized and so demanding that to keep abreast of the literature in even a narrow field of research occupies an uncomfortably large share of one's leisure. How then is it possible to find time to master the skills and background of other disciplines? Thirdly, many scientists are introverts and many societies are inclined to be exclusive, although I hasten to exclude biochemists and The Biochemical Society from this generalization, because many biochemists are as extrovert as the average salesman—well almost—and this Society has been particularly fortunate in that many of its Officers have been dynamic personalities who have made their influence felt far outside the boundaries of biochemistry and indeed outside the boundaries of the United Kingdom. In practice, however, most scientists do not have that self-assurance in argument or aggressive advocacy of a particular point of view that commercial life requires.

"Yet I believe that industry needs scientists in management and on boards of directors and I think that scientists should be made aware of the openings that exist and of the qualities and qualifications required of those chosen to fill these vacancies. . . ."

Robinson also contributed very interesting remarks on the 'economics of research' and 'forecasting the return from research projects' and finally discussed the 'recruitment of managers from scientific staff'.

F. C. Webb's arresting title was 'From microgram to hundredweight'. After considering 'experimental training and industrial practice—a contrast in scale', he advocated special courses designed for industrial application and gave examples in biochemical engineering.

A. T. James of Unilever talked about the 'Relationship between industrial and academic laboratories'. Displaying candour and insight about the present position he noted a large degree of misunderstanding concerning the aims as well as the facilities of some of the larger industrial laboratories. He went on to suggest:

1. Interchange of staff on a short term basis both for acquiring specialized techniques and participating in teaching, on the one hand, and industrial research on the other. Industrial scientists could often give short courses on special subjects and add to the teaching expertize of the university department. Such teaching should be recognized by the granting of some special status to the lecturers.
2. Industrial laboratories of acceptable standard should be recognized as suitable for study for higher degrees. The necessary supervision could be provided.
3. University departments should arrange joint colloquia with their industrial colleagues when there was sufficient common interest.
4. Universities seeking financial assistance from industry for research projects would find the process easier and more effective when the organizations had a formal link.

James was impressed by the difficulties experienced in recruiting large numbers of first class scientists for the development side of research.

On re-reading the account of this symposium it is clear that biochemists were already making positive suggestions on lines which have since been suggested by many other persons and organizations, and to some extent put into effect.

RELATIONS WITH OTHER BODIES

CHEMICAL SOCIETY LIBRARY

When the Biochemical Society was formed there was no early likelihood that it could have a permanent office or a library. Many members also belonged to the Chemical Society, the Library of which received support from the Chemical Council which negotiated the method by which different societies subscribed to the upkeep of the Library.

The Biochemical Society was invited in 1919 to participate in a scheme whereby the members would be allowed to use the Library in Burlington House under almost exactly the same conditions as Fellows of the Chemical Society. Copies of the *Biochemical Journal* were presented to the Chemical Society and an annual donation made towards the Library. Until this time many requests for an exchange of the *Biochemical Journal* for that of another Society had been made but had not proved feasible. Exchanges now became possible and with the consent of the Library Committee (one member of the Biochemical Society serving on it) exchanges were made with journals that would be a new addition to the Library. A happy and fruitful system of co-operation grew up.

The Society's subscription was initially £10 per annum and for some years it remained at that level. Soon after the 1939–1945 War expenditure on books and journals began to rise phenomenally and the Society agreed to take a share in the maintenance of the Library. By 1955 this share slightly exceeded £300 and by 1962 it was over £900.

A new basis of assessment was decided upon in 1963. The net maintenance costs of the Library were to be shared on the basis of the membership of the contributing societies with an allowance for overlap which had been calculated in 1961. Until that time the assessment had been calculated on the previous year's costs and not on the current year's costs. This meant that there was always a deficit which had been met by the Chemical Council but whose funds for this purpose were now running out. In 1964 the Society's contribution was £1,080 and for the financial year ending December 1968 it was £1,454.

There is no doubt that this arrangement has been very valuable to members of the Society.

The Chemical Council has had a long history of service. In 1963 various societies which published original research benefited by grants from the endowment fund of the Chemical Council. The Society at that time obtained 21 per cent of £5,678, that is, £1,192. It was decided in 1965 to wind up the Chemical Council but no decision was taken about distributing the assets. This was a slow process but in 1968 a draft scheme for winding up the Council had been prepared by the Charity Commissioners, and the Endowment Fund of the Council was duly transferred to the trusteeship of the Chemical Society.

BIOCHEMICAL ABSTRACTS

The Society's concern with information services was evident as far back as the year 1925 and subsequent developments illustrate how difficult it has become to organize dissemination and retrieval systems, particularly when, as in biochemistry, information comes from varied sources.

In 1927 the Committee considered a suggestion that members might purchase *British Chemical Abstracts* "A" without joining either the Chemical Society or the Society of Chemical Industry. Many members were buying Americal *Chemical Abstracts* at £2. 2. 0d. per annum! A glance at the footage of shelving occupied today by one year's *Abstracts* reveals the almost runaway growth of the literature of Chemistry. The situation was not however in 1927 so serious as to cause acute anxiety.

In October 1934 C. R. Harington proposed that the Society should contribute £100 towards *British Chemical Abstracts*. The costs had up to then been borne largely by the Chemical Society but the position had been reached that 23 per cent of the cost was for biochemical abstracts and the Society should accept some responsibility. At that time many biochemists still belonged to both Societies and there was opposition to the proposal. Nevertheless the principle was accepted. A year later the Chemical Society appointed a committee to consider *Biochemical Abstracts* "A" (Pure Science) as a matter of urgency and A. C. Chibnall served as the Society's representative. The outcome was a request that representatives of the Chemical, Biochemical and Physiological Societies and of *Nutrition Abstracts* should discuss the future of *Biochemical Abstracts* and eliminate unnecessary duplication. A joint Committee with J. B. Leathes as Chairman advocated extending the Bureau of Chemical Abstracts to form a Bureau of Chemical and Physiological Abstracts to cover the field embraced by *Chemical Abstracts, Physiological Abstracts* and the abstracting journals supported by the Imperial Agricultural Bureaux Council (*Nutrition Abstracts and Reviews*). The abstracts would fall into sections, the appropriate section to be available on preferential terms to members of the constituent organizations. Late in 1936 it emerged that a combined effort to publish one comprehensive set of Biochemical and Physiological Abstracts would be regarded favourably by the Chemical, Physiological and Biochemical Societies but not at that time by the publishers of *Nutrition Abstracts and Reviews*. The Chemical Society was willing to continue to spend £1,800 per annum on the production of *Biochemical*

Abstracts while annual support at the rate of not less than £300 was available from the Physiological Society and at the rate of not less than £200 from the Biochemical Society. The joint committee recommended that the independent publication of *Physiological Abstracts* should cease and the Bureau of Chemical Abstracts be reconstituted on the lines suggested.

As an interim measure the actions taken worked well. In 1939 the A III Section of *British Chemical and Physiological Abstracts* received £1,500 from the Chemical Society, £300 from the Physiological Society and £200 from the Biochemical Society but this was not nearly enough. Harington persuaded the Committee to increase the Society's annual contribution to £300 for three years. The decline in volume of scientific literature during the war relieved the situation a little. In 1943 it was decided that as from January 1944 there should be a separate C section entitled 'Analytical Methods and Apparatus' the extra expense to be divided equally between the Chemical Society, the Society of Chemical Industry and the 'A III' fund. About this time Harington resigned from the Bureau and F. G. Young took his place. The Committee placed on record their great appreciation of Harington's services.

In 1944 the Bureau of Chemical and Physiological Abstracts became incorporated, with The Biochemical Society as a constituent body and F. G. Young as a Director. J. H. Bushill, R. L. Synge and R. K. Callow in turn represented the Society. Soon after the war the Bureau offered *Abstracts* A III to members of the Society for £1 per annum plus 5/– for *Abstracts* C and it was possible to obtain them printed on one side of the paper.

The A III *Abstracts*, edited by Samson Wright, were on the whole greatly valued although they were sometimes criticised as too concise. There was, however, continued anxiety over cost of production, and the Chemical Society became restive under its share of the burden. By 1950 Callow had to report that the finances of the Bureau were in a precarious state and the Society agreed to increase its contribution, but by 1952 the Bureau had gone into voluntary liquidation. The Royal Society had recently been asked by the Lord President of the Council to discuss a number of matters concerned with financing chemical publications. The Committee of the Society approved a memorandum on *British Abstracts* Section A III drawn up by Callow and others and forwarded to the Royal Society. The document stated that

> "The Biochemical Society considers that the issue of abstracts in the subjects covered by Section A III is of service to Biochemistry and is anxious to co-operate with the Physiological Society in exploring the possibility of continuing publication.

"The Biochemical Society would be prepared to continue for the present its financial contribution to such abstracts at its present rate while obtaining abstracts for its members at a reduced price related to the cost of production. Unfortunately the Society is, in common with others, finding the cost of publication of original papers a considerable strain".

In August 1953 it was decided that the Bureau should cease to produce *Abstracts* after completing the Indexes for that year. Other arrangements were to be made to continue *Abstracts* A III and the Society of Public Analysts would continue Section C. A new journal called *British Biological Abstracts* was proposed and The Biochemical Society agreed to provide £400 for 1954 and be ready to add another £600 if required.

Early in 1954 a company limited by guarantee and to be known as Biological and Medical Abstracts Ltd was formed to publish *Biological and Medical Abstracts* or *British Abstracts of Medical Sciences*. The publication did not however have a big circulation and failed to pay its way. It was reported in June 1955 that Biological and Medical Abstracts Ltd had agreed to accept an offer made by Pergamon Press Ltd to assume responsibility for publishing *British Abstracts of Medical Science*. The Committee of the Society agreed to continue its financial support at the rate of £400 for 1955. From July 1956 the publication was renamed *International Abstracts of Biological Sciences*.

In 1958 the price of (American) *Chemical Abstracts* to university departments had reached $80 per annum and *International Abstracts* approximately £30 per annum. Negotiations between *Biological Abstracts* and *International Abstracts of Biological Sciences* were initiated in 1957 at the suggestion of the International Council of Scientific Unions but did not go well. It was expected that 24,000 papers would be abstracted in 1959 but the circulation of the new journal remained too small. The arrangement whereby members of the Society received the biochemistry section of the *Abstracts* at a preferential rate was continued but in 1960 the price was raised to £1. 12. 6d. with £1. 15. 0d. for the index. The Society continued its payment of £400. In 1960 the *Abstracts*, in four volumes, covered nearly 25,000 papers. By September 1961 the increase in papers necessitated a further increased charge to £2 per annum. Nevertheless in June 1962 the total loss on the publication had reached £17,500 of which £12,000 had been written off by Pergamon Press. The price of the *Abstracts* was raised by £10.

In July 1963 the Committee noted that the number of members buying the Biochemistry Section of the *Abstracts* had fallen to 310 and it was decided not to continue the payment of £400 to the publication. It was agreed that those members who wished to purchase the Biochemistry and Biophysics Section of the *Abstracts* should pay the full price.

In 1964 it was reported that the 'break even' circulation was still far away.

In some respects the story of *Abstracts* is a melancholy one. The Society steered a middle course but the magnitude of the expansion in papers and reports has been enormous. A generation or two ago it was possible to inspect the literature rather closely and the individual scientist could just about afford to buy, bind and find shelf space for abstracts in his field. Today he relies greatly on serial publications presenting accounts of recent progress on selected problems but he still uses abstracting journals for a bout of hard work in a library.

Many distinguished members of the Society have given much service over *Abstracts*, have borne considerable burdens and have had to make difficult decisions. It may be doubted whether anybody could have done much better under constantly changing circumstances.

There has been a decline in the role of abstracting journals and a not unrelated failure to discover earlier work. Work done 40 years ago was often unnecessarily repeated 20 years later and today the same topics are being investigated again, apparently because it is "easier" to repeat the work than to comb the literature.

Today there is a widespread realization that the publication avalanche has made necessary a new approach and the Chemical Society in association with other Societies (including the Biochemical Society) has formed a Consortium on Chemical Information (COCI). This body drew up in July 1968 its Terms of Reference and outlined a programme. Members of the Biochemical Society will be much interested in this venture.

BIOLOGICAL COUNCIL

Towards the end of the 1939–1945 War some kind of grouping or federation of biological societies was widely felt to be desirable and particular emphasis was placed on 'co-ordination of publication'. Scientists indeed foresaw rapid growth in papers and articles submitted for publication but there is little evidence that the size of the avalanche was correctly anticipated. The Biochemical Society took the initiative in exploring the need for a Biological Council.

A questionnaire (July 1944) raised many issues and was followed by a meeting held in the rooms of The Royal Society at Burlington House in September 1944. The Society's comments were in favour of rationalization of production (printing, format, indexing, etc.) and it recommended co-operation which was "particularly needed to channel papers to the right journal". This was thought to be best left to editors "but there was a continual over-riding of boundaries . . . . resulting in the founding of new journals to serve particular fields". There was an interesting interplay of ideas and the Committee thought it very undesirable "that natural development impulses should be artificially inhibited. Any attempt to formulate rigid divisions at the present time would be inexpedient in that it would tend to inhibit the development of the future".

W. T. J. Morgan, F. G. Young and R. A. Peters were active in forming the Council and the latter represented the Society on the Committee of Formation. Supporting societies were asked to contribute "£5 per annum or £1. 1. 0d. per 100 members, whichever sum be the lesser", but by 1947 the subscription was raised to £10.

In 1948 the Biological Council considered establishing an Institute of Biology on the lines of the Royal Institute of Chemistry. The Committee of the Biochemical Society treated the proposal rather cautiously at first, but in 1949 sent a donation of £5 towards the Institute of Biology then being set up.

The British National Committee for Biology sponsored by The Royal Society was more concerned with international relations.

Very little reference to the Biological Council appeared in the Minutes of the Committee of the Society for many years. The Institute of Biology was of course successfully launched and there is now some overlap between its membership and that of the Biochemical Society. The Institute has its own publications and its special roles, one of which under N. W. Pirie is to keep a look out for words, new or old, in need of definition.

PARLIAMENTARY AND SCIENTIFIC COMMITTEE

The Parliamentary and Scientific Committee provides liaison between scientific societies and members of both Houses of Parliament. In 1947 the Society paid a subscription of £15. 15. 0d. and was allotted three representatives. The principal subject for discussion in 1948 was an import quota for scientific books. The President of the Board of Trade announced in the House of Commons that the import quota of learned scientific and technical books had been increased from 100 per cent to 200 per cent by value of pre-war imports.

The Parliamentary and Scientific Committee advocated an improved supply of paper for scientific books and journals, and a memorandum was issued seeking information about shortage of technological and scientific staff. Biochemists did not then seem to be experiencing a shortage of applicants for junior technician posts and an acute shortage of graduates was not expected. There was however anxiety about the further training of non-graduate laboratory assistants.

The Society's subscription to the Committe was raised to 25 guineas in 1954 and over the years a succession of representatives attended meetings without (so far as the records show) having much to report.

After serving for five years a representative of the Society reported that in his opinion the Parliamentary and Scientific Committee did offer a channel of approach to appropriate ministers but the performance of the Committee fell far short of its real potential and radical reorganization was needed. There was probably no connexion between these strictures and the fact that in 1966 A. C. Frazer and R. A. Morton gave talks on problems concerning food additives to the Committee! The report, however, of the Society's new representative was that while there was no doubt about the timeliness and interest of the talks, the rather haphazard contact between members of Parliament and scientific organizations was not a sufficient link between scientists and politicians.

To be fair there are many links and if some are not publicised that may not necessarily be a disadvantage.

BRITISH ASSOCIATION FOR THE ADVANCEMENT OF SCIENCE

In 1956 the Society suggested to the British Association that a section on biochemistry should be established. The Association decided to invite Professor F. G. Young to give an evening discourse on 'The growth of biochemistry'. A little later the Association agreed that the name of Section I, Physiology, should be changed to Physiology and Biochemistry. This suggestion had the full support of the Section of Physiology, and bore fruit quickly and in particular the Cardiff meeting of the Association in 1960 resulted in a very good joint meeting.

The Society decided to co-operate with the Association in finding biochemists to address sixth forms at schools.

The Manchester meeting of the British Association also displayed biochemistry and physiology together. This new development is particularly associated with the name of K. S. Dodgson and the role of biochemistry in the British Association is now fully established in the section for biomedical sciences.

JOINT MEETING WITH BRITISH ASSOCIATION

A meeting held jointly with Section I (Physiology/Biochemistry) of the British Association for the Advancement of Science at Cardiff on 7 September 1960 marked a departure from the usual scientific activites of the Society. There were nine speakers, all concerned with the organization and financing of biochemical research in this country. An overall picture emerged of the development of a rapidly advancing scientific discipline and a Report bearing the title *The Organization and Financing of Research in Biochemistry and Allied Sciences in Great Britain*, edited by K. S. Dodgson, was published in 1960. In an introduction, Sir Alexander Haddow said that British biochemistry had made "innumerable contributions at fantastically low expense" and quoted a tribute by Vannevar Bush to British scientists for their "extraordinary ability to get important scientific results with a minimum of expensive equipment"; adding, however, "that they sometimes made things difficult for themselves by under-equipment". He also quoted H. Melville "whether we like it or not, the age of string and sealing wax has come to an end, and the pursuit of scientific knowledge is now a rather expensive and complex affair in which substantial resources in men, material and facilities are needed." Sir Alexander drew attention to an imbalance between expenditure on the life sciences and that on physical, engineering and military sciences, but in his view the root of the problem lay not so much in expenditure on research as on scientific and other education.

F. J. C. Herrald discussed the broad policy of the Medical Research Council and described the rôle of biochemistry in Units and Institutes. Sir William Slater, dealing with the Agricultural Research Council, paid special attention to the support of biochemistry at the Institute of Animal Physiology at Babraham and at the Rowett Institute in Aberdeen. He was much concerned over recruitment of biochemists for agricultural research. C. Jolliffe considered the 'Rôle of the Department of Scientific and Industrial Research' particularly with regard to studentships, fellowships and research grants, and outlined what was then thought to be the future policy of the Department of Scientific and Industrial Research (now the Science Research Council). 'The Imperial Cancer Research Fund' was dealt with by G. F. Marrian who described the work of his laboratory and discussed wider aspects of the financing of cancer research with its larger voluntary aspect.

F. C. Happold dealt with university departments of biochemistry. Starting from the purposes of a university he went on to discuss finding money for rising running costs and new equipment. This led him to compare the rôles of the Research Councils and the University Grants Committee and to emphasize the importance of grant-aided postgraduate work with reasonable technical assistance.

J. B. Bateman, (then Deputy Scientific Attaché at the American Embassy, London) speaking about 'American Support of Biochemical Research in Great Britain', referred to the scale of support from private as well as from Government sources. It was however not easy to identify purely biochemical projects. The situation at the time (1960) was that the United States sponsored biological and medical research in the United Kingdom to the extent of some 5–6 per cent of the total British Government non-military allocation. "Although a certain amount of this goes into 'pure' biochemistry, the support tends to be spread over a large number of inter-disciplinary projects; it is obvious that nearly all of these have their biochemical aspects and in many cases they do in fact involve participation of a biochemist." In support of basic research in Great Britain, the

United States of America "placed some emphasis upon the care taken to ensure that this sponsorship cannot be construed as an attempt to interfere with local patterns of research".

A. Neuberger, commenting on Sir Alexander Haddow's remarks on the inadequate supply of biochemists in this country, related this to lack of sufficient information about biochemistry in schools. He advocated elasticity in university appointments and doubted whether the training of biochemists for industry was adequate.

THE FOUNDING OF THE NUTRITION SOCIETY

In 1940 E. V. McCollum in his *History of Nutrition* thought with some reason that the chemistry of vitamins had probably been completed. It is interesting that in Britain the first informal meetings of nutrition workers which led in time to the formation of the Nutrition Society began in 1940. There was a feeling that Nutrition needed a degree of independence, shown in England by "informal meetings of Nutrition workers" organized by S. K. Kon.

"In 1941 after certain preliminary conclaves and discussions, a manifesto, drafted by Sir John Orr (later Lord Boyd Orr), was circulated, signed by the following eleven Heads of Institutes (eight of them members of the Biochemical Society):

Sir John Barcroft (Chairman, Food Investigation Board)
Dr Harriette Chick (Head of Division of Nutrition, Lister Institute)
Professor J. C. Drummond (Professor of Biochemistry, University College London, and Scientific Adviser to the Ministry of Food)
Dr J. Hammond (Physiologist, Animal Research Institute, Cambridge)
Dr L. J. Harris (Director, Dunn Nutritional Laboratory, Cambridge)
Sir Frederick G. Hopkins (Professor of Biochemistry, Cambridge)
Professor H. D. Kay (Director, National Institute for Research in Dairying)
Sir Charles Martin (late Director, Lister Institute)
Sir Edward Mellanby (Secretary, Medical Research Council)
Sir John Boyd Orr (Director, Rowett Research Institute)
Professor R. A. Peters (Professor of Biochemistry, Oxford)

"This manifesto stated . . . 'there are a considerable number of research workers and others in favour of holding meetings to discuss nutritional problems. . . . If there is a sufficient number of workers who wish to hold meetings for discussion of nutritional problems, the best procedure would be to form a society on the lines of the Physiological and the Biochemical Societies although there would be no question of publishing a journal in the meantime'. (Actually it was only a very short period before this "question of publishing a journal" did in fact come to the fore!)

"The next step, in July 1941, was a conference, held in London at the Royal Institution, presided over by Orr, at which representatives from twelve different centres of research were present. "(L. J. Harris *Brief Note on the History of the Nutrition Society* Nutrition Soc. 1963).

An inaugural conference was held in Cambridge in October that year and the address was given by F. G. Hopkins (his last public appearance).

A Scottish group of the Society was formed and it has flourished ever since. The membership was at first about 75 and by the time of the 100th meeting in 1956 it was 750; it is now 1152.

The Society began to publish its proceedings in 1944 and the *British Journal of Nutrition* came a little later. S. K. Kon was editor for many years and left his imprint on the *Journal* and the Society.

In 1946 the Society organized with the help of the British Council what was in effect the first International Congress of Nutrition. It was largely a European venture and lasted 17 days, which included visits to many centres. The Congress led to the foundation of the International Union of Nutritional Sciences (IUNS) with L. J. Harris as first Secretary-General (from 1946 to 1960). This body has organized Congresses in London, Basle, Amsterdam, Paris, Washington, Edinburgh and Hamburg. It is affiliated officially to the United Nations organizations, for example, FAO, WHO, UNESCO and the Council for International Organizations of Medical Sciences.

In 1963 Harris wrote a short history of the International Union of Nutritional Sciences starting from a meeting in July 1946 with Sir Joseph Barcroft in the chair. An International Provisional Committee met in London in June 1946 and a constitution was drawn up. Recognition came quickly except that negotiations with ICSU were long drawn out. This seems to have hampered the International Union very little.

Relations between the Biochemical Society and the Nutrition Society have been harmonious and there has of course been considerable overlapping in membership. A few persons have served—at different times—on both editorial boards. The *Journal of the Science of Food and Agriculture*, published since by the Society of Chemical Industry now fills another niche in the publication of papers embracing *inter alia* a wide field of applied biochemistry. This journal also includes an excellent abstracts section. It can be said that the Societies concerned with disseminating information about biochemistry, nutrition and food science have reached an equilibrium that is not unsatisfactory.

In the course of time many other new Societies with a biochemical side have been formed and no doubt there will be others. If the Nutrition Society can be fairly regarded as a daughter Society, the formation of Groups within the Biochemical Society can be seen as a cohesive influence!

INTERNATIONAL AFFAIRS

FIRST INTERNATIONAL CONGRESS OF BIOCHEMISTRY

An International Physiological Congress was held at Oxford in 1947 and The Biochemical Society approached the organizers "to ensure that Biochemistry was allocated its share of the programme". Their reply said that it was "impossible to issue a general invitation to biochemists to participate in the Congress and that while no actual embargo would be placed on biochemical papers, these would have to come from, or be introduced by, members of the Physiological Society". This was felt to be a rather dusty answer and the Committee of The Biochemical Society thereupon took the first steps to promote an International Congress of Biochemistry. The Committee of the Physiological Society confirmed their stance but added that "if the Biochemical Society decided to initiate Congresses of their own they would have the Physiological Society's blessing, encouragement and offer of assistance"!

Preliminary enquiries addressed to French and American biochemists did not arouse marked enthusiasm for International Congresses and a circular sent to 100 persons elicited only 24 replies but these were all in favour. The Committee was not however deterred. It asked E. Baldwin to explore the feasibility of holding a Congress at Cambridge in August 1949. The Colleges agreed that Cambridge could accommodate at least 500 people "provided that the food situation was not too difficult". A little later the Colleges agreed that they could find places for 780 people, "including a few women".

Early in 1948 a large and influential Congress Committee was set up, and given major responsibility. The Society set aside £1,500 for a Congress Guarantee Fund. The arrangements proved very successful and the Congress was attended by 1,700 people of whom about 700 came from 32 overseas countries. A.C.Chibnall was appointed President and in his opening address (19 August 1949), he reviewed the growth of biochemistry and explained how, for want of an International Union of Biochemistry, the Biochemical Society had taken the initiative. The Congress was officially recognized by the International Union of Chemistry "without commitment for the future" and a meeting of accredited representatives of various countries under the chairmanship of Sir Charles Harington "would discuss arrangements for organizing future Congresses." The President recalled that eight congresses had been organized by the Société de Chimie Biologique since 1927 and many biochemists from outside France had attended.

Chibnall went on to explain that a number of friends and colleagues of F. Gowland Hopkins had produced a commemorative volume made up of a collection of his writings and appreciations of his work. The Congress Committee took pleasure in presenting a copy of the volume to every member attending the Congress. Cambridge scientists were proud of Hopkins's great achievements.

Canon C. E. Raven, Vice-Chancellor of the University (himself a naturalist and author of a well-known biography of John Ray) conferred degrees of Sc.D. *honoris causa* on C. F. Cori, Viscount Addison, K. Linderström-Lang, Sir Charles Harington, J. Trefouel and A. W. K. Tiselius. A Garden Party given at St John's College (20 August 1949) was much enjoyed.

In addition to the programme of papers on a wide range of topics, three Congress Lectures were delivered, the first by C. F. Cori entitled 'Influence of hormones on enzymatic reactions', the second by M. Florkin on 'Biochemical aspects of some biological concepts' and the third by Sir Robert Robinson on 'Tryptophan and its structural relatives'.

The final session of the Congress was held in the Senate House on 25 August. Harington announced the names of the members of his committee and reported a resolution to accept an invitation from the Société de Chimie Biologique to hold the next International Congress of Biochemistry in Paris in 1952. The International Committee for Biochemistry would consist of 20 members, with power to co-opt. It was further decided that the new Committee "be instructed to approach the International Council of Scientific Unions with a request for recognition as the international body representative of biochemistry, with a view to the formal constitution of an International Union of Biochemistry as soon as possible". The resolutions were carried by acclamation.

Complimentary speeches about the Congress were made by K. Linderström-Lang, A. J. Kluyver, D. van Slyke, J. Roche and A. W. K. Tiselius. Miss V. Moyle then presented to Professor Chibnall an album containing over 1,700 signatures of members. The President expressed his thanks and spoke with appreciation of the work of Colonel Griffin as Honorary Organizer, and of F. Dickens and E. Baldwin as Honorary Secretaries. Valuable support had been given by E. C. Dodds and C. R. Harington.

Early in 1950 the Society agreed to publish a Report on the opening and concluding Sessions and to include the three Congress Lectures in full. Consisting of 50 pages with four plates it was printed for the Biochemical Society and made available at the price of "3s. net". This very modest price sharply illustrates the monetary inflation that has since occurred. There was actually a small surplus of about £500 on the running of the Congress.

It was natural that a first International Congress of Biochemistry, being held at Cambridge, should worthily commemorate Gowland Hopkins as a man and as a savant. This did not however prevent the Congress from being essentially forward-looking. There was something in the air, especially in informal exchanges, to suggest that after the lean years of the war biochemistry had made a fresh start. It was seen as an independent discipline with exciting cross-links with other subjects. There was moreover a pervasive sense that the subject had begun to fuse chemistry and biology with recognizable effects in increasing understanding and promoting the general welfare. Nevertheless, after the grim experiences of war, biochemists kept their feet on the ground. Despite the misgivings felt in 1947 about food rationing, the college catering staffs managed very well under far from easy conditions.

The prevailing themes of private conversations showed how world food problems could not be ignored, how the same or related problems were shared by different countries, how there was so much to be done, so many students to be taught and research workers to be trained and so much money needing to be raised in competition with other necessary projects. The mood was one of restrained optimism.

After the successful first International Congress of Biochemistry at Cambridge further congresses have been held in other capital cities. Each congress has involved a vast amount of work by the local organizing committee and the resulting publications have increased in size year by year.

So far as British participation is concerned the officials of the Society have been very successful in raising money, firstly from The Royal Society as administering Government grants and subsequently from The Wellcome Trust which has been particularly generous. Substantial amounts have also been provided by a number of industrial firms. Altogether the Society had at its disposal for the Moscow congress £5,000 for the Tokyo congress no less than a sum of £10,000 was obtained from The Royal Society, The Wellcome Trust and The Biochemical Society. The total amount raised was approximately £11,100.

On the way home from Tokyo, 70 visitors, mostly from the United Kingdom, attended a special meeting at the Indian Institute of Science at Bangalore.

Many participants in these international congresses have expressed the opinion that there was an optimum size for scientific meetings and that it had been exceeded. Indeed it was in mind at one stage that a maximum of

4,000 persons should be imposed but it should be added immediately that such a proposal arose originally as a local response to accommodation difficulties, and lack of facilities. There was no decision of principle and it was never clarified, whether the 4,000 included wives who might be attending the social functions but not necessarily the scientific meetings. Be that as it may, the problem of whether an international biochemical congress can be too successful remains on the whole a matter deserving further study, both from the point of view of overall expense and success judged on scientific and professional criteria. In passing it may be of interest to note that at the time of writing consideration was being given to the organization of a new type of congress on at least one occasion, in that plans were in hand for three related conferences to be held simultaneously in three different Swiss towns in 1970 to replace an international conference, plans for which had to be abandoned in June 1969.

THE INTERNATIONAL UNION OF BIOCHEMISTRY

It was widely realized that international congresses of biochemistry would be facilitated by the formation of an International Union of Biochemistry and its recognition by the International Committee of Scientific Unions (ICSU).

But the project soon ran into heavy weather. Early in 1949 the (British) National Committee for Chemistry had resolved "that the proposal for an International Union of Biochemistry would be better replaced by a proposal to establish a joint committee between the International Union of Biological Sciences and the International Union of Chemistry which should be its mother union". The Congress Committee (set up at the Cambridge meeting) advised the Committee of the Biochemical Society to make further soundings at home and abroad.

In September 1949 the International Union of Pure and Applied Chemistry (IUPAC), meeting at Amsterdam, decided to reconstitute itself in six sections one of which would be devoted to Biological Chemistry under the Chairmanship of Professor A. W. K. Tiselius. The Biochemical Society's Committee however adhered strongly to its view that Biochemistry should be represented internationally by a separate International Union of Biochemistry (IUB). J. N. Davidson reported that the proposal had also been discussed by the British National Committee for Biology but support was withheld "mainly on the grounds that a multiplicity of unions was to be deplored". It was pointed out that "a mixed commission of the Unions of Chemistry and Biology serving the needs of Biochemistry would be virtually independent of both parent bodies and would draw its funds from the International Council and not from the parent organizations."

The International Union of Pure and Applied Chemistry grew out of an International Association of National Chemical Societies founded towards the end of the 19th Century. By 1967 IUPAC had held 24 conferences and had acquired considerable influence through its various Divisions, Sections and Commissions. It must be agreed that IUPAC in 1949 had already shown a quite considerable interest in biochemistry. In consequence a substantial number of scientists greeted the proposal to form an International Union of Biochemistry with open hostility, others, doubting the need, quietly withheld support. The unofficial International Committee for Biochemistry was set up at the Cambridge Congress and held its first meeting in 1949. The second meeting (Copenhagen 1950) attended by J. N. Davidson, reported that draft Statutes for an International Union of Biochemistry had been sent to the International Council of Scientific Unions but had elicited very little sympathy for the proposed new union. In fact the IUB did not formally come into existence until 1955.

The Biological Chemistry Section of IUPAC continues to do good work (see C. E. Dalgliesh, *Chemistry & Industry*, 4 May 1963) although the main congress-organizing function in the field of biochemistry has been transferred amicably to the IUB. Various international symposia have however been held under the auspices of IUPAC [for example, Peptides, Munich 1949; Pharmaceutical Chemistry, Florence 1962; Clinical Enzymes, Ghent 1962; Macromolecular Chemistry, Toronto 1968, with the following planned; Steroids and Terpenes, Mexico City 1969; Clinical Chemistry, Geneva 1969].

There is now a Joint Co-ordinating Committee of the two Unions. 'This committee provides for the interpretation to IUB of the thinking, interests and actions of IUPAC and is responsible for ascertaining and interpreting to IUPAC the thinking, interests and aspirations of the biochemists.' It tries to ensure that congresses and symposia organized by both Unions are complementary and not competitive, and it aims to avoid clashes in dates.

The Biological Chemistry Section has organized commissions on Nomenclature of Biological Chemistry, on Proteins and on Clinical Chemistry (since 1951). The President of the Commission on Nomenclature is W. Klyne who succeeded Byron Riegel. Excellent work has been done on problems connected with fat-soluble vitamins, steroids, coenzymes, vitamin B_{12} and quinones with polyisoprenoid side-chains. The Commission has collaborated closely with an IUB Commission of Editors of Biochemical Journals. The Clinical Chemistry Commission has done fine work (under the presidency of E. J. King and more recently M. E. Freeman). It has organized International Congresses of Clinical Chemistry (Amsterdam 1954, New York 1956, Stockholm 1957, Edinburgh 1960, Detroit 1963) and a joint sub-commission with the IUB Enzyme Commission has operated successfully.

The Biological Chemistry Section of IUPAC thus performs very useful functions. It numbers among its former presidents Murray Luck, E. C. Dodds, A. W. K. Tiselius, E. J. King and W. Sperry.

We may now return to the formation of IUB. In October 1950 the National Committee for Chemistry drew the Society's attention to the XII International Congress of Pure and Applied Chemistry to be held in New York in September 1951. E. C. Dodds had been nominated as a delegate. He was then invited also to represent the Biochemical Society. About that time the Committee of the Society approved draft statutes drawn up by C. R. Harington for the proposed IUB and appealed to The Royal Society to set up a National Committee for Biochemistry.

By March 1951 an official application to establish an International Union of Biochemistry had been made to The International Council of Scientific Unions and was due to be considered at a meeting in Washington in October 1951. F. Dickens and J. N. Davidson were invited by the Committee to travel to Washington as spokesmen for the Biochemical Society. The arguments in favour were to be set out in a memorandum to be agreed by Davidson, Dickens, Dodds and Harington.

When IUPAC met at New York in September 1951, Tiselius became its President, leaving the presidency of the Biological Chemistry Section vacant. E. C. Dodds was asked at very short notice to accept nomination as president of the Section. He agreed with some misgiving "and only on condition that it was understood and minuted that he was in favour of an independent Union of Biochemistry and that he would continue to further the cause of an independent Union". The envoys sent by the Society to the Washington meeting felt that at the time many American biochemists were apathetic about international Unions. Murray Luck had however written to Harington to invite the International Committee (set up at the Cambridge Congress) to nominate five persons to fill vacancies on the Biological Chemistry Section Committee of IUPAC.

There can be no doubt the whole IUB project suffered a set-back in September 1951 and Dodds's acceptance of the Presidency of the Biological Chemistry Section met with some criticism. One of those who served on the Committee at this critical time recalls a sense of dismay. "Although Dodds continued to work for a separate International Union of Biochemistry some members felt that the battle might have been lost. When, however, R. A. Peters became Committee Chairman in 1952 he at once made it clear that he would press for a separate Union whatever the opposition and this firm stand restored flagging confidence at home and rallied international opinion". The Committee of the Biochemical Society were daunted but not defeated. It was decided to provide opportunities for discussion and to invite members to make their views known.

In May 1952 a questionnaire on the proposed IUB was sent to members of the Society. The number of replies (479) was somewhat disappointing but a large majority of them favoured continuing to press for an independent Union. There was also a large majority in favour of instructing the International Committee to establish 'in the meantime' a working arrangement with the IUPAC Section of Biological Chemistry.

In June 1952 the Society's committee heard from Davidson that American opinion among biochemists was now veering towards an independent Union. The Committee again decided to work to that end but raised no objection to any British representatives on the International Committee serving on the IUPAC Section Committee if invited, "but they should continue to press for an independent Union".

The third meeting of the International Committee of Biochemistry was held in Paris in July 1952 during the second International Congress of Biochemistry. In the absence of Harington, Davidson took the Chair. The International Committee approved the action taken at Washington. However, the Biological Chemistry Section of IUPAC called for a joint meeting with the International Committee. Davidson, again in the Chair, records that the proceedings were stormy but the International Committee for Biochemistry held its ground.

In September 1952 the Executive Board of ICSU met at Amsterdam and the International Committee for Biochemistry (represented by M. Florkin (Belgium), E. Brand (U.S.A.), H. G. K. Westenbrink (Holland) and J. N. Davidson (Great Britain) put the case for IUB. No final decision was however reached but "as a result of unofficial advice given to us during the meeting, but outside it, the International Union of Biochemistry was in fact established independent of ICSU."

By 1953 an Interim Council had been set up and applications to become members of an International Union of Biochemistry were being considered. The Biological Chemistry Section of IUPAC met in Stockholm in the summer of 1953 with Dodds in the Chair. On this occasion the proposed IUB gained the definite support of the Section and the Secretary of the Biochemical Society "was instructed (by the Committee) to thank Professor Dodds for his valuable work". The Society then approached The Royal Society to set up a National Committee for Biochemistry. This however, could not appropriately be done before admission of an International Union of Biochemistry had been considered at the next

General Assembly of The International Council of Scientific Unions in 1955. In the meantime the Biochemical Society decided to act as the interim adhering body and to set up a provisional National Committee.

A meeting of the Constitutive Assembly of IUB (which had developed from Harington's International Committee) was held in London in January 1955 and a formal request for recognition by The International Council of Scientific Unions was sent forward then so as to give the necessary six months notice. The visiting delegates were entertained at receptions organized by the Biochemical Society, by the University of London and by the Ciba Foundation.

The British National Committee for Biochemistry was formally set up by The Royal Society in November 1955 under the Chairmanship of Sir Rudolph Peters.

The IUB did not come into being at all easily and the delay was perhaps necessary so that opinion could ripen. In retrospect the struggle was worth while and few will doubt that the symbiosis between IUPAC and IUB corresponds to a certain inter-relationship between chemistry and biochemistry. It may be pertinent to remark that biochemistry presents so many problems that no single Society can—or even should now try to cope with them all. This may be illustrated by examples which happen to be familiar to the writer. IUPAC has published in pamphlet form (Butterworth's Scientific Publications) Reports on 'The vitamin D bio-assay of oils and concentrates'; 'The vitamin A potency of β-carotene'; 'The assay of vitamin A oils' and 'The determination of copper content of foodstuffs by a photometric method'. Another set of reports concerned methods for the determination of toxic substances in air. The problems were tackled by representatives of 'pure' and 'applied' biochemistry when the time was judged to be ripe. IUPAC already had great experience in planning similar exercises in international co-operation.

There is still a place for symposia under the auspices of IUPAC particularly on inter-disciplinary topics (for example, steroids and terpenes at Mexico City 1969, on clinical chemistry, Geneva 1969). In analytical chemistry much effort has been devoted recently to a provisional method for the determination of aflotoxin.

The arguments for continuing some at least of IUPAC's work in the biochemical field were as convincing as those in favour of IUB and an amicable arrangement for co-existence became necessary. Nowhere was this more desirable than in discussions about nomenclature. (see p. 113)

EUROPEAN BIOCHEMICAL SOCIETIES

The Biochemical Society played a leading part in the formation of the Federation of European Biochemical Societies. Perhaps it would be truer to say that a small group of enthusiasts piloted the Society, with considerable skill, into taking the initiative on lines that were destined to gain general acceptance.

Joint meetings with a few European societies were held in the fifties, for example a meeting at Turku in Finland had proved very successful, although the cheaper travel rates skilfully obtained by the Officers of the Biochemical Society meant flying in the small hours to Copenhagen, and a long wait before continuing to Stockholm where a second long interval had to be endured. For the Cambridge meeting of the Society in 1960 special invitations had been extended to members resident on the Continent of Europe and the response was encouraging. A joint meeting with the Dutch Biochemical Society was held in 1963 and about 100 people from the United Kingdom attended. A colloquium, on the broad topic of plant biochemistry, was very well attended.

By 1961 the Committee had decided that the Society's activities were expanding so rapidly that an additional honorary secretary was needed. In March 1962, F. C. Happold proposed (subject to approval of a change in the Rules at the Annual General Meeting) that W. J. Whelan should be appointed to the newly created office of International Secretary. It is interesting that the Committee regarded this title with more favour than that of Foreign Secretary. Whelan held the post until December 1966 and A. P. Mathias succeeded him in 1967. The appointment of Whelan as International Secretary was a clear indication that the Committee of the Society was ready for European co-operation.

Whelan wrote to the Secretaries of the various biochemical societies in Europe suggesting a meeting in a different capital each year. He also asked for details of all the meetings of the societies so as to prepare a joint calendar.

An international meeting was planned for Oxford in 1963 and the Wellcome Trust agreed to grant £450 towards defraying travel expenses of invited speakers from European countries. Contributors from the U.S.A. found that travelling expenses could be met from their own research funds. The Oxford meeting was one of the largest ever held by the Society and in addition to the Symposium nearly 120 papers were read. The Meeting Dinner had to be split between two Colleges.

It was arranged that at this meeting discussions should take place with representatives of European societies (the Biochemical Society being represented by Happold [Chairman of Committee] and Whelan), the outcome being that it was hoped to inaugurate the Federation at the beginning of 1964. Draft statutes were discussed and it was agreed the Federation of European Biochemical Society meetings should replace the Biochemical Society's international meetings.

By November 1963 the Federation of European Biochemical Societies had been adhered to by 13 countries with others about to join. The first formal meeting of the Council was planned for March 1964 in London.

A full meeting of the Federation took place at University College (23–25 March) with a registration fee of 25/– intended to make the meeting financially independent of the Society. It was planned to publish the papers read at the meeting separately from the *Biochemical Journal*. The Society, however, agreed to produce Abstracts and circulate copies. The Symposium, on the structure and activity of enzymes, was to be published as a book. This first meeting of the Federation was recognized as a special occasion and Lord Boyd was good enough to invite the Federation delegates, the invited speakers, the Society's committee and others to dinner at the Goldsmiths' Hall on 24 March. The Society's Committee decided that the Symposium should be published by the Academic Press as No.1 in the Federation of European Biochemical Societies Symposia Series, organized by the Biochemical Society.

In May 1964, S. P. Datta (Honorary Treasurer, FEBS) reported a loss on the London meeting of £1,700. An economic registration fee would have been £3 but it was thought that sales of the Symposium volume and Abstracts might yield £500. The FEBS committee desired a levy of 1/– per member per annum and the Biochemical Society agreed to pay up to £200 per annum.

The second meeting under the auspicies of FEBS was arranged for Vienna in 1965. A small summer school on ultracentrifugation was arranged at Louvain and directed by C. R. M. J. de Duve. This lasted 10 days (8–18 June 1965), the registration fee was £25, and the school was attended by 31 members, six from the United Kingdom. The Vienna meeting was held in April 1965 and 11 out of 55 applicants received travel grants through the Biochemical Society, aggregating £700. At Vienna a FEBS subcommittee was set up to consider launching a new journal of biochemistry. A new post, Secretary-General of FEBS, was created and Whelan was appointed the first holder of the office.

The subcommittee decided to recommend setting up a *European Journal of Biochemistry* with a representative editorial board. It favoured finding a commercial publisher and laid down conditions for starting negotiations. The Committee of the Biochemical Society discussed the proposals at length but finally decided to support the project. When it became known that some journals might abandon publication in favour of the new Journal, the Committee of the Biochemical Society hastened to declare that it had no interest in abandoning publication of the *Biochemical Journal!* Shortly afterwards it became known that the German Biochemical Society had agreed that *Biochemische Zeitschrift* should cease publication and that the new European Journal would take its place. An arrangement was made with Springer Verlag to sell the new Journal to members of constituent societies for £7 per annum and £23 per annum to others, postage included. The new Journal got off to a good start with 1500 institutional and 200 private subscribers.

The third FEBS meeting was held at Warsaw (4–7 April 1966) and about 90 persons from Britain attended. The fourth FEBS meeting was held at Oslo (3–7 July 1967) and the Biochemical Society found travel awards totalling £700. The fifth FEBS meeting was held in Prague on 15–20 July 1968 and the sixth meeting in Madrid (7–11 April 1969) and a substantial number of travel grants were awarded.

The idea of summer schools proved attractive to many. At Glasgow a school on steroid biochemistry was held in April 1967. The Committee of the Society met a deficit of £50, "but only because the U.K. was on this occasion the host country". Nevertheless the Committee later agreed to contribute £50 towards a FEBS course on computer techniques in biochemistry at Edinburgh. A further FEBS summer school was held at Rehovoth in September 1968 the subject being nucleic acids and since then several other gatherings have been arranged.

Whelan retired as Secretary-General of FEBS early in 1967 and was succeeded by H. R. V. Arnstein who was proposed by the Norwegian Biochemical Society and seconded by the Biochemical Society.

It soon became necessary to increase the FEBS subscriptions from 1/– to 2/– per member per annum and this was generally accepted. The next proposal was to issue a new journal called *FEBS Letters*. The project was considered by the Committee of the Society and the Minutes for 20 September 1967 laconically record that support was given "after considerable discussion". This new FEBS venture is described as "an international journal which offers the most rapid possible publication of papers from the broad fields of biochemistry, biophysics and molecular biology which are short, complete and essentially final reports. The over-riding criteria for acceptance are novelty and general interest. In addition *FEBS Letters* contain brief authoritative reviews of special fields which are developing rapidly". The first issue appeared in July 1968 "and publication is at least monthly". If *FEBS Letters* match the exacting prospectus they will have a bright future!

FEBS publications now include accounts of meetings (Abstracts) and of symposia. Four volumes covering colloquia held at the Vienna meeting have appeared (Pergamon Press) and three volumes covering symposia at the Warsaw meeting are available (Academic Press or Polish Scientific Publishers, Warsaw, jointly). The Oslo meeting produced a volume of Abstracts and six volumes of Proceedings published jointly by the Oslo University Press and Academic Press.

There can now be no doubt whatever that the time was ripe for the sort of development which has given rise to FEBS and its publications. It is true that meetings, travelling expenses and publications have for some time been increasing exponentially and that the expansion must eventually level off despite the phenomenal growth of biochemistry as a subject for undergraduate and postgraduate studies at universities. The individual scientist is facing an avalanche if not a plethora of printed matter and it is not easy to decide what selection principles he should apply when it comes to buying and housing literature for his personal use, not to mention reading it. New problems may well arise but the enterprise of a few members of the Society, especially Whelan and Happold, certainly met a response from the officers of the European societies and the astonishing progress made by FEBS is a tribute to efficiency and enthusiasm in a combined operation of considerable magnitude.

AUSTRALIAN BIOCHEMICAL SOCIETY

In 1954 Dr R. Lemberg wrote to the Society to say that it had been decided to form an Australian Biochemical Society. A letter was sent conveying good wishes to the new Society. The first meeting was held in May 1956 and Lemberg (retiring President) wrote to say that V. M. Trikojus had been elected President and F. D. Collins Honorary Secretary. In 1966 it was felt that closer liaison with London was desirable and in February 1967 the Australian Society expressed a desire to have a guest lecturer from England periodically. It proved a little difficult to complete appropriate arrangements but eventually the first visit was undertaken by B. S. Hartley in 1969, and since biochemists in New Zealand had also shown an interest in these arrangements the tour was extended to cover both Australia and New Zealand. Financial assistance for the project was forthcoming from the Royal Society, with the Biochemical Society making its own contribution, and it is hoped that with this type of support that the occasion may become an annual event.

NOMENCLATURE, NOTATION AND ABBREVIATIONS

Every Society that publishes scientific papers has to cope with ticklish problems concerning communication. The Biochemical Society has had a full share of activity and responsibility over these matters. On some occasions the Committee discussed particular situations but over the years much of the work has been done at the level of the Editorial Board. The Society as a whole is greatly indebted to a comparatively small number of chemists and biochemists who have given time and thought to problems which everybody recognizes to be important but few are able and willing to tackle.

The machinery now in operation is working well and it is worth while to trace something of its recent history.

In September 1923 A. Harden was invited to preside over a IUPAC Committee on Biochemical Nomenclature. The Committee of the Society nominated W. M. Bayliss and J. A. Gardner to represent the Society and the Chemical and Physiological Societies each appointed a member. The next reference in the Minutes dealt with a report presented to the Committee by J. C. Drummond in 1925, who was apparently somewhat perturbed. His experience at annual conferences on nomenclature led him to make the following recommendations:

"1. Close co-operation between biochemists in Great Britain and America be to established.
2. In both countries this co-operation to extend also to workers in organic chemistry.
3. A report to be prepared by these interests, representing English speaking chemists, stating their views relating to:—
    (a) Biochemical Nomenclature generally.
    (b) The changes proposed in the resolution passed at recent meetings of the International Union of Pure and Applied Chemistry.
    (c) The question of the constitution of the International Union particularly as to whether it is a properly constituted and competent authority to decide matters relating to Biochemical Nomenclature."

In December 1925 the Minutes state that "Dr Sherman of Columbia University outlined in a letter the steps which he proposed to take in America with a view to dealing with the question of Biochemical Nomenclature. Harden's subcommittee was asked to meet and requested to take such action as it thought fit . . . and to exchange views with Dr Sherman".

In February 1928 a report from the subcommittee was sent to the Accessory Food Factors Committee. [This body, set up jointly by the

Medical Research Council and the Lister Institute of Preventive Medicine, was the main national committee on vitamins]. The Committee of the Society supported the subcommittee's recommendations with regard to the terms vitamin B_1 and B_2 and added that explanatory adjectives such as antineuritic and antipellagra be added.... The Committee recommended that the Editors of the *Journal* "be requested to urge authors to adopt this nomenclature". In May 1928 Harden reported that 'certain' American workers considered that the decision of the Committee with regard to this matter was somewhat peremptory. The Secretary was asked "to write to the Secretary of the Accessory Food Factors Committee to say that the present working arrangement was not to be regarded as other than provisional".

In its way this incident is instructive. Today IUPAC publications are careful to refer to 'proposed rules, tentative rules and informal discussions' but the caution shown is less a sign of weakness than of hard experience.

An interesting problem arose in 1955 when the Society for Analytical Chemistry, through its Biological Methods Group, invited the Society to send two representatives to an informal meeting to "discuss the desirability of standardizing statistical nomenclature". It was agreed that R. B. Fisher and W. O. Kermack should be asked to represent the Society but whether they were able to do so is not on record in the Minutes.

Another problem arose in 1956 when E. Lester Smith wrote to say that he, K. Folkers and K. Bernhauer, attending a vitamin B_{12} symposium at Hamburg, would "endeavour to reach some agreement on the nomenclature of compounds allied to vitamin B_{12}". The Committee of the Society agreed that the project had its goodwill but suggested that the outcome of the discussion at Hamburg should be communicated to the International Union of Biochemistry and the Commission on Nomenclature of IUPAC. A year later A. G. Ogston as Chairman of the Editorial Board reported that the findings had been sent to IUPAC but he advised that "the recommendations should not be adopted by the *Journal* until they had been ratified by IUPAC."

In 1958 W. V. Thorpe and E. J. King reported on a meeting of Editors of Biochemical Journals held during the Vienna International Congress. There was considerable willingness "to attempt to establish uniformity in nomenclature, abbreviations, etc. It was probable that a Commission would be set up by the International Union of Biochemistry to deal with the matter."

For many years the Biochemical Society had made available a printed document giving instructions to authors on the presentation of papers and the use of abbreviations and units. Periodically revised after searching and sometimes trying discussions, the document has played a very useful part in reducing friction and maintaining consistency in the *Journal*.*

The International Union of Pure and Applied Chemistry has on the cover of Information Bulletin Number 32 (August 1968) the inscription "50 YEARS IUPAC. 1918–1968." Reference has already been made to opposition when the formation of the International Union of Biochemistry (IUB) was under discussion, and of course the record of the Division of Biological Chemistry of IUPAC was a cause of pride. The work done in Biochemical Nomenclature was outstanding.

At this stage it is of interest to quote from *Comptes Rendus XXIV Conference IUPAC* (Prague, September 1967) ratifying the decisions of an informal discussion held in London in April 1967.

Minute 15. *Relationships with IUB.*

The President referred to the history of the relations between IUPAC and IUB with a view to bringing these two bodies fully into co-operation with one another; the Executive Committee had authorized him with the Secretary General to meet and to discuss with Professor S. Ochoa, President of IUB, the best means of achievement. Details of a proposed agreement between the two Unions were set out in the Minutes of the Executive Committee and of the meeting with the President of IUB dated 28 April 1967. In the discussion which followed, Professor W. M. Sperry gave a historical review of the activities of the Division of Biological Chemistry; he was in general agreement with the proposal although he deplored the disappearance of the Division of Biological Chemistry which it entailed. The Council agreed unanimously that the proposals outlined by the President offered the possibility for a close and friendly relationship which would be in the best interests of both Unions and of chemistry as a whole. Council therefore
Resolved:
  (a) That the Division of Biological Chemistry in IUPAC be dissolved.
  (b) The Joint IUB/IUPAC Nomenclature Commission be continued as part of the Division of Organic Chemistry.
  (c) That Clinical Chemistry be made a Section of the Division of Analytical Chemistry.
  (d) The co-ordination of activities of the two Unions be achieved by cross-representation at the IUPAC Bureau level.
  (e) That implementation of the above resolutions be dependent on their acceptance in principle by IUB at its General Assembly in 1967.

Minute 16. *Clinical Chemistry*

The growing importance and scope of clinical chemistry brought about in part by development of new analytical methods had led to the creation by Members of the Commission on Clinical Chemistry

---

*The latest edition, **Biochemical Journal** (1969) **112**, 1, is entitled 'Policy of the Journal and Instructions to Authors.' The copies can be obtained from the Executive Secretary price 2/6.(U.S.A. $0.30)

of the International Federation for Clinical Chemistry. The Federation having grown up in association with IUPAC was anxious to see the association maintained in the future and welcomed the suggestions made in Minute 15 that the present Commission of Clinical Chemistry should become a Section of the IUPAC Division of Analytical Chemistry. The case for continuing association, which had been put to the Executive Committee, was outlined by Dr. M. Sanz and Council.

Resolved

That the International Federation of Clinical Chemistry be granted the status of an associated organization of IUPAC.

The decisions reached were eminently satisfactory from the standpoint of nomenclature.

A great deal of good work had already been done. The IUPAC–IUB Commission on Biochemical Nomenclature consists of 10 ordinary members, five of whom are nominated by IUPAC and five by IUB and one corresponding member. The present Chairman comes from IUB whereas the Secretary is a representative of IUPAC. Thus, the Commission is a true liaison group working under the auspices of both Unions. This status is quite recent; until 1 January 1964, there had been an IUPAC Commission for the Nomenclature of Biological Chemistry (founded 1921) which then was transformed to the present set-up. The Commission meets once every year for three days. Meetings took place in 1964 in Anif (Austria) and in 1965 in Paris and the "Work of the Commission" was summarized as follows:

1. "Abbreviations and Symbols for Chemical Names of Special Interest in Biological Chemistry". Tentative rules under this title were compiled by the mentioned former IUPAC—Information Bulletin No. 20, 1963; a new revised edition is in preparation. The rules on abbreviations and symbols have been adopted by most existing journals of biochemistry throughout the world.

2. "Rules for the nomenclature of various classes of substances:
   (a) Trivial Names of Miscellaneous Compounds of Importance in Biochemistry (will come out in Information Bulletin 25, 1965, has already come out in *Biochim.biophys.Acta*, **107**, 1–4, 1965).
   (b) Nomenclature of Quinones with Isoprenoid Side-Chains (will come out in *Biochim.biophys.Acta*, **107**, 5–10, 1965).
   (c) Nomenclature and Symbols for Folic Acid and Related Compounds (has come out in Information Bulletin 23, 1965, and in *Biochim.biophys.Acta*, **107**, 11–13, 1965).
   (d) Abbreviated Designation of Amino Acid Derivatives and Polypeptides (submitted for publication).
   (e) Nomenclature of Corrinoids (final draft ready).

3. "Sub-Commissions are working on the following problems:
   (a) Nomenclature of lipids.
   (b) Nomenclature of synthetic polypeptides.
   (c) Nomenclature of cyclitols.

4. "A Sub-Commission for a revision of the nomenclature of carotenoids will be newly founded in the next few months.
5. "The Commission is represented by two of its members in the Committee on Carbohydrate Nomenclature, founded by the IUPAC Commission on Organic Nomenclature."

Excellent relations exist with commissions working in neighbouring fields and all documents of the Biochemical Nomenclature Commission are submitted for examination to the IUPAC Commission of Organic Nomenclature and to the IUB Commission of Editors of Biochemical Journals before being published. Many members of the Society have contributed to the work on nomenclature none perhaps more than W. Klyne and E. C. Slater. On the Interdivisional Committee on Nomenclature of IUPAC the Biochemical Commission is represented by its Chairman.

Information Bulletin Number 25 (February 1966) contains a general statement from the Commission on Biochemical Nomenclature signed by O. Hoffman-Ostenhoff as Chairman. There is an interesting report on trivial names of miscellaneous compounds of importance in Biochemistry. It deals with vitamins A_1, B_1, B_2, nicotinamide, C, D, E and with isoprenoid quinones including vitamins K, plastoquinones and ubiquinones and related substances. There is also a section on the abbreviated designation of amino acid derivatives and polypeptides. A report from the combined IUPAC–IUB Commission (Information Bulletin Number 30, October 1967) appeared "for discussion" on the nomenclature of lipids, including sphingolipids and neuraminic acid. The August 1968 Information Bulletin (Number 32) included tentative rules for a one-letter notation for amino acid sequences and also tentative rules for cyclitol nomenclature. The same Number includes the draft of an extensively revised version of the 1959 'Manual of physicochemical symbols and terminology', with an important exposition of the International System of Units (SI units).

Information Bulletin Number 31 (March 1968) gives the membership of Commissions on Trace Substances, Food Additives and Contaminants and Mycotoxins. Biochemists will also be interested in a provisional IUPAC method for determining aflotoxin.

The Report of the 1967 discussions (S. Ochoa, W. Klemm and R. Morf) referring to the joint Commission, ended as follows:

"These measures are intended to provide a satisfactory atmosphere for fruitful and close co-operation between the two Unions:
(a) As a first step in that direction a joint IUB/IUPAC—Ciba Symposium on Stereochemistry be held in 1968.
(b) That the two Unions should act concertedly in education and in influencing editors of biochemical journals with a view to ironing out and co-ordinating the pattern of publishing scientific results.

(c) That the two Unions coordinate their efforts in effecting cooperation between the abstracting services in chemistry and biological abstracts."

As has been said (p. 105) the Biochemical Society took the initiative in founding IUB. Problems inevitably arose but there is now general satisfaction with the harmonious relations which have been firmly established.

PATENTS

At different times matters concerning patents have come before the Committee of the Society but no agreement seems ever to have been reached.

In 1929 a subcommittee met Dr F. H. Carr who on that occasion represented the Association of British Chemical Manufacturers. The question at issue was whether the Biochemical Society could approve a report on Patent Law Reform which already commanded substantial agreement among the bodies making up the Federal Council of Chemistry. The subcommittee was divided and asked the Chairman to call a special meeting of the full Committee "as an important matter of general principle was involved". The Committee turned out to be deeply divided, all shades of opinion being held. The memorandum favoured by the Federal Council of Chemistry included a Scheme of Dedicated Patents. H. W. Dudley felt reluctant to continue as the Biochemical Society's representative on the Federal Council. This gave rise to a full discussion at the end of which Dudley formally enquired "whether he were empowered to inform the Council that the Biochemical Society Committee approved the Scheme of Dedicated Patents." The answer was in the negative by five votes to three. Dudley then resigned as the Society's representative on the Council and the Chairman of the Committee was asked to take the vacant place. The divided opinions in 1929 reflected wider dissension about the ethics of patents relating to natural products. One school of thought for instance regarded vitamin research as a co-operative effort in which patentable steps and unpatentable ones could be equally significant. Another school thought that the abnegation of patent protection of almost every aspect of natural products necessary for health was neither justifiable nor practical.

It seems from the Minutes that the topic was dropped. Perhaps discretion was really the better part of valour; the Society could have been split down the middle at that juncture.

In 1944 the then President of the Board of Trade appointed a Departmental Committee on Patent Law. The Association of British Chemical Manufacturers decided to set up a joint Chemical Committee on Patents to prepare and submit evidence on behalf of Chemistry. When the Committee of the Biochemical Society was invited to nominate a representative it took the line that the Society could not be committed to any definite course of action without any specific point or points being discussed by the Committee. The Chairman (at that time E. C. Dodds) was then appointed to the Joint Chemical Committee. He later reported that the matters discussed were "largely technical or legal". He had signed the memorandum in a personal capacity and had not committed the Society in any way.

Much water had by that time flowed under the bridges. The chemotherapeutic revolution had occurred and war-time collaboration with the Americans had occurred over penicillin. Various schemes of dedicated patents had proved successful in the U.S.A. Public opinion among scientists had changed, from the Medical Research Council outwards. It was no longer possible to sustain sharp distinctions between either synthetic and natural substances or between medicinals and other biologically serviceable compounds. Today the National Research Development Corporation includes biochemists on its staff and the situation over patents is no longer a source of controversy on matters of fundamental principle.

INFORMATION EXCHANGE GROUPS

Many biochemists have been concerned with a very interesting experiment—or group of experiments on 'Information Exchange Groups'. Originally set up and financed by the National Institutes of Health in the U.S.A. the description favoured by the sponsors was "a continuing international congress by mail". Members of a group were sent copies of typescript reports of the very latest results obtained by other members. Circulation to members of the group was not formally a 'publication'. By July 1965 seven information exchange groups were in existence, the largest, dealing with oxidative phosphorylation and electron transport (under the Chairmanship of D. E. Green) had 450 members while another on nucleic acid and the genetic code (under J. D. Watson and M. Nirenberg) had 350 members. The groups attracted both support and hostility. It was made clear by the National Institutes of Health that "any scientists actively engaged in research in any one of the seven research areas was eligible for membership". There were no dues. The National Institutes of Health were servicing the groups as a pilot project.

The conception of 'Information Exchange Groups' was defended by D. E. Green (*Science*, 1964, **143**, 308) and the objections were neatly noted and countered. In July 1965 the Committee of the Society discussed 'I.E.Gs' using an admirably succinct and fair assessment prepared for them by H. R. V. Arnstein, P. N. Campbell and R. R. Porter. Any arrangement whereby new results or opinions could be made known within a few days to several hundred workers in the same field, without being subject to normal scrutiny by referees or editors was bound to make as many enemies as friends. There was a fear that unfinished work might be circulated prematurely. Some scientists were dubious about considerations of priority. Did the circulation by an I.E.G. constitute a 'publication' or not?

In 1966 the International Commission of Editors of Biochemical Journals discussed future policy in relation to the publication of papers which had already been published (presumably in abbreviated form) by an Information Exchange Group. W. V. Thorpe, as Secretary of the International Commission, consulted by post the members of the Editorial Board of the *Biochemical Journal*. The Board failed to approve the recommendations of the International Commission of Editors and sharp differences of opinion clearly existed. In December 1966 the Committee of the Society heard that Information Exchange Groups "in their present form would cease after 21 March 1967" although some kind of modified scheme might eventually take their place. A relieved Committee referred the problem of I.E.G.s to the Advisory Committee for Publications.

Reaction to Information Exchange Groups was partly a matter of temperament and partly of environment. Some, who thought that research was in danger of becoming too competitive and publication too frequent and too hasty, saw them as exacerbating an already unhealthy state of affairs. Others who saw research as driving irresistibly on and wished to remain in the main stream valued the I.E.G. as preventing them from getting into a back water. Whether they would have paid economic dues for the service is another question altogether.

CELEBRATIONS

Although the Biochemical Society has prided itself on its informality and never had a President or a Mace it has always been ready to have celebrations.

The One Hundredth Meeting coincided with the Annual General Meeting on 13 March 1926 and a dinner was held at the Grosvenor Hotel with F. Gowland Hopkins in the Chair. No record of the speeches is known to survive but the signatures of those at the dinner were collected on a large card and framed. A reproduction appeared in Plimmer's History and the original is in the Department of Biochemistry at University College London.

Twenty-first Birthday. To mark this anniversary a dinner was arranged on 17 November 1933 at the Hyde Park Hotel "at a cost not exceeding 12/6d. with gratuities"; A. Harden, J. A. Gardner and R. H. A. Plimmer were honoured; 158 members and guests under the Chairmanship of Gowland Hopkins, then President of The Royal Society, attended. The well known photograph of original members was taken at the 21st anniversary meeting. (See also Meetings, p. 73).

Nobel Laureates. Four distinguished scientists, C. Eijkman, F. G. Hopkins, H. von Euler and A. Harden were awarded the Nobel Prize in 1929. This notable event could not be allowed to pass without recognition. With the exception of Professor Eijkman, who was not well enough to travel from Utrecht, the Laureates were entertained at dinner on 3 February 1930 at the Hotel Victoria. The President of The Royal Society, the Presidents of other societies, and a representative of the Physiological Society, were all invited as guests. The Swedish Minister honoured the Society by his presence. Over 200 members and guests were present. A card bearing signatures of those present was framed and presented to the Lister Institute for custody.

FIFTIETH ANNIVERSARY MEETING

The official date of the founding of the Biochemical Society is 4 March 1911. The Annual General Meeting in March 1961 therefore provided a very suitable occasion for celebrations. The writer happened to be Chairman of Committee from 1959–1961 and the enthusiasm with which the Golden Jubilee celebrations were organized is a happy memory. The Committee was greatly assisted by a host of well wishers, by many former officers and by the authorities at University College London.

The Biochemical Society.
(FOUNDED MARCH 4TH. 1911)

The Hundredth Meeting Celebration Dinner.
Grosvenor Hotel, Victoria, S.W.
MARCH 13TH. 1926.

[Signatures of attendees, including:]

H. S. Raper, Percival Hartley, F. Gowland Hopkins, J. J. Irving, R. A. Peters, Mary Dudley, C. Rimington, H. Chaussen, Barbara Wytcham-George, Olga Hartley, E. A. Harrison, H. W. Dudley, R. D. Lawrence, Raymond Hartley, Elsie Schryver, H. W. Kinnersley, Ed. Hinks, G. F. Marrian, H. Gordon Reeves, E. Walker, G. L. Peskett, Dorothy B. Strabbe, M. Boas, J. A. Gardner, K. Culhane, Muriel Bond, V. B. Reader, Hilda A. Channon, M. H. Drummond, J. C. Drummond, J. S. Yeates, Harold Kirg, Ethel King, E. Horton, Beatrice Kay, H. D. Kay, J. H. Quiltam, H. J. Coombs, T. S. Hele, Nan Gluckstein, Edith P. Lampitt, Sue Elworthy, J. M. Duncan Scott, R. J. Buchheimer, Hilda W. Slater, W. K. Slater, S. Kossilov, Ray Robinson, J. B. S. Haldane, M. Fleet, N. E. Gowland Hopkins, Robert Robinson, J. C. Hoet, S. Hughes, J. Jephcott, M. Jephcott, Phelia Collets, E. R. Keugh, V. G. Plimmer, M. Cook, John Pryde, V. K. Shipston, Maud Hyne, E. A. Armstrong, Harriette Chick, G. T. Majorban, K. Tansley, W. Charles Chap., H. Roaf, R. K. Cannon, Franic L. Gram, Nora Armstrong, J. Rosedale, E. C. Liddy, G. Macleod, Leslie J. Harris, Hugh Maclean, Ernest H. Starling, E. Armstrong, S. G. Hedley, E. B. Verrilie, Florence Starling, R. H. A. Plimmer, Arthur Harden, Jas. F. Weddell, Beatrice Roaf, E. Baldwin, Marjory Stephenson, Dorothy Jordan Lloyd, George Barger, Paul Haas, W. H. Burch, J. P. Fielding, A. C. Chibnall, E. Price, Henry E. Armstrong, Cecil Chibnall, Hugh Smith, J. Amslongh, J. Pirie, G. A. V. Lunde, Temple Gray, D. J. Hayon, Emil Wiedenhall, John Lowndes, Helen R. Downes, J. M. Tunhull, Henry J. Ellis, W. S. Paterson, I. Banerji, E. J. Brinkman, Al. Bachrach, Hans Lam, K. Grace Blaney, P. Exploton, Margaret Bech, Mary S. Smith, C. H. Best, Laura Evans, C. Lovatt Evans

Signatures of members present at the Hundredth Meeting Celebration Dinner
"Presented to the Biochemical Department, University College London
by the Committee of the Biochemical Society"

CELEBRATIONS

The fiftieth Anniversary Meeting was held in London from 27–29 March 1961. The proceedings began on 27 March with a symposium on 'The structure and biosynthesis of macromolecules'. The meeting was held in the Beveridge Hall, Senate House, and was extremely well attended. The first session, under the Chairmanship of F. G. Young included a paper by P. Doty (Harvard) and the second session, presided over by W. T. J. Morgan, included papers by J. N. Davidson, E. L. Hirst and W. Z. Hassid (Berkeley). The following day A. Neuberger was in the Chair and papers were presented by M. F. Perutz, E. L. Smith (Salt Lake City) with A. Light and J. R. Rimmel, and by J. Monod (Paris).

On 28 March H. A. Krebs delivered the Third Hopkins Memorial Lecture at The Royal Institution, Albermarle Street, W.1. The historic lecture theatre was filled to overflowing and many were unable to gain admission. Krebs spoke on 'The physiological role of ketone bodies'. He took the occasion to recall the circumstances under which he settled in England as a refugee from Nazi persecution and his moving tribute to the friendship and assistance accorded to him by Gowland Hopkins will long be remembered by all who had the privilege of listening to it.

During the evening of 27 March a Conversazione was held in the North Cloisters at University College London and many friendships were begun and old ones renewed. No less than 650 people attended the Conversazione. Substantial donations towards the cost were made by Glaxo Laboratories, The Distillers Company, Messrs Kemball, Bishop & Co. and Messrs John & E. Sturge & Co.

On Wednesday 29 March, 34 original communications were presented in three different lecture theatres between 10 a.m. and 1 p.m. The Annual General Business Meeting was held in the Anatomy Theatre at University College and J. N. Davidson became the new Chairman of Committee.

The Anniversary Dinner was held in the New Refectory, University College and the Society was honoured by the presence of a very large number of distinguished visitors. The principal guest was Lord Hailsham, then Minister of Science. The toast of 'The Guests' was proposed by J. N. Davidson. Speaking earlier in Glasgow, Lord Hailsham had emphasized the need for the universities to provide more graduates and especially more scientists. Davidson agreed that there was a shortage of biochemists, but although there were many vacancies for fairly junior 'tame' biochemists there were not enough top-level posts and some universities were still without biochemistry departments. Among the guests were the Presidents and representatives of a long list of scientific societies, headed by Sir Howard Florey, P.R.S. The Royal Society had recently celebrated its three hundredth anniversary. Davidson drew attention to the fact that the Jubilee of the Society coincided with the centenary of the birth of Gowland

Hopkins, a founder member of the Society, an early Chairman of Committee (1913–1914) and a former President of the Royal Society.

Representatives of nearly 20 sister biochemical societies were welcomed. At the time of the Jubilee celebrations there were three and only three Honorary Members of the Society, Sir Rudolph Peters, Sir Henry Dale and Sir Charles Harington. All three were happily able to be present. Reference was made to Sir Rudolph's great sevices as President of ICSU (International Council of Scientific Unions) and to Sir Charles's long editorship of the *Journal*. Two original members, Sir Charles Lovatt Evans and G. W. Ellis were also present.

Lord Hailsham, Professor Florkin and Academician Oparin replied on behalf of the Guests. Sir Howard Florey proposed 'The Society' and R. A. Morton replied. The speeches at the dinner were reported in *Chemistry and Industry* (22 April 1961) and the account is reproduced here by permission.

BIOCHEMICAL SOCIETY

50th ANNIVERSARY DINNER

LONDON, MARCH 29, 1961

"The Biochemical Society held a dinner on 29 March to celebrate its 50th Anniversary. The Guest of Honour was the Rt Honourable Viscount Hailsham, Q.C.,* Minister for Science, and the toast was proposed by Sir Howard Florey, President† of the Royal Society.

"Replying to the toast of the guests, Lord Hailsham said that when the Biochemical Society was founded 50 years ago, biochemistry was a new discipline on the fringe of two others better established, and was seriously in danger of falling between two stools. After 50 years this could hardly be claimed to be the case. Biochemistry was on the map. In terms of membership, the Society's numbers had doubled in ten years passing the 3000 mark, thus becoming the largest society in the world devoted solely to biochemistry.

"The growth of biochemistry over those 50 years was an illustration with a general moral in the development of science. At first sight science was no longer a single discipline. In one sense there was nothing in common between the young engineer who last year was awarded the D.S.I.R.‡ Wolfe prize for his hydrostatic transmission, and the Cambridge

---

* Now Mr. Quintin Hogg.
† Later Lord Florey.
‡ Now Science Research Council (S.R.C.)

scientists who unravelled the molecular structure of haemoglobin, or the M.R.C. scientists who were working on Interferon, or the C.E.R.N. physicists with their 'strange' particles, or the geologists of South Kensington. The difference between these disciplines was sufficient to destroy the myth of any general 'policy for science' determinable at any given time by a suppositious scientific planning board.

ORGANIC WHOLE

"In another, and more intimate sense, science was an organic whole. No sooner did two of its disciplines, like physics and chemistry, or chemistry and biology, or biology and mathematics, or chemistry and engineering, grow so far apart that they were made necessarily separate studies, than cross-fertilization took place between each pair: nuclear physics rendered possible new diagnoses in medicine; electronics opened up new paths in biology. It was not possible in the long run to differentiate the almost biological growth of science—except at the cost of killing the parts. Of all this, the Biochemical Society was at first a pioneer, and now a classical exemplar.

"This was reflected in the field of public expenditure on science. In terms of Government support, about a quarter of the Medical Research Council units were primarily or solely biochemical; about 10 per cent of the Agricultural Research Council's spending was on pure or applied biochemistry; and, of course, an immense amount of biochemical work was also undertaken in the course of investigations not primarily biochemical in character. In addition to the two biological Councils the D.S.I.R. had current at least 45 grants in the field of biochemistry, although it was doubtful whether the Universities were even now grasping the full extent of the help that D.S.I.R. could give them.

"Biochemistry thus had become a main theme in scientific research on the biological side, and, although occasionally one heard lamentations that Government aid to it was too small, there was no chance in the foreseeable future of biochemistry being able, any more than any other scientific discipline, to by-pass the requirement that to qualify for Government aid a scientist must first run the gauntlet of a jury of his brother scientists on behalf of one of the Research Councils or the U.G.C.

SCIENTIFIC PLANNING

"It was important that not merely the structure, but also the philosophy, of our present arrangements should be brought home to scientists. In all modern states it was necessary that the scientific planning should be done by scientists—not some specially-recruited scientific civil service so much as the men and women who were actually doing the work, wherever they were—in the universities, in industry, in Government institutions, and in the professions. It was not a question whether Government ought to do it. The fact is that Government could not do it—as was shown by experience everywhere, whether in the Soviet system where it was done through the Academy, or in the American system where it was done through a Foundation and a sort of extension of the President's private office, or here in Britain where it was done through the Research Councils and the Atomic Energy Authority under the Minister of Science. There was no magic in the particular structure adopted. Each structure was largely a commentary, not on scientists or science, but on the political constitution of the state concerned. This was the real argument against any fully-fledged "Ministry for Science". It existed nowhere in the world, and anyone who reflected on the nature of the problem must see it to be a chimera.

"Lord Hailsham said the function of Government was not to do the science, but to see that science was being done, and was being applied. Even this it could only do for the most part indirectly, by encouragement of science in the Universities, by stimulation of science in industry, and by the organization of institutes for research in the several fields where, if it was not done directly in this way, it would not be done at all. It was extremely important not to dogmatise about this. The general pattern was extremely varied, and corresponded largely to the complexities of the economic and social life it served.

"The growing point for pure science of all sorts in this country ought to be, and mainly was, in the universities. It was sometimes said that here it ought to be entrusted to the University Grants Committee. Four years' experience, however, showed that our present system was better. The University Grants Committee was concerned with the permanent work and structure of Universities. The introduction of new research was better handled separately, and this was the function of the Councils. Even in the last four years the D.S.I.R. had increased its grant-giving function by something like a factor of three. And it was not only D.S.I.R. For example, some of the most distinguished research work, like that on the haemoglobin molecule in the Cavendish, was done on Medical Research Council money. This separate source of finance for new research projects was a real guarantee against undue conservatism, and both the newest disciplines, like molecular biology, and the still relatively new, like biochemistry, stood to gain by this arrangement.

"It was, however, in industry that many of the applications of science, including biochemistry, had to be found. Here the pattern was far more patchy. The industries perhaps closest to biochemistry—notably the pharmaceutical and fermentation industries—had a fairly good story to tell. But in some of the older industries it seemed probable that a radical change of outlook was needed towards the role of science and scientists in industry.

"In the meantime there was a definite part which only scientists—and particular sorts of scientists—could do for themselves. It was often said, for example, that there were not enough places for biochemists in our Universities. This might, of course, be true. But against it must be balanced the equal and opposite truth that ,if there were more demand for places, more places would in fact be provided. This meant, of course, that, apart from any other means of development, there was room to do even more to sell the idea of biochemistry to students as a means of creating a demand and thus also the provision of grants—and ultimately the provision of research grants as well.

"The toast of the Biochemical Society was proposed by Sir Howard Florey, P.R.S., who began by referring to his great admiration for biochemists. It was quite clear, he said, that practically all the problems in medical and biological research would, in the last instance, have to be solved by biochemists, or possibly biophysicists. He considered the growth of biochemistry in the 20th Century to be the outstanding event in experimental biological science.

"He had not the slightest doubt that the Biochemical Society, from the time it was founded, had a great influence on bringing biochemists together for their discussions and meetings, and it must have had a profound influence on the teaching of, and research into, biochemistry in the universities. Today, of course, it was a great Society with a very large membership. The fact that there were so many guests from abroad attested its international renown, and there was no doubt that the *Biochemical Journal* was one of the ornaments of scientific literature.

"The Society since its founding had continued to exert a great influence; this could not be better attested than by the extremely successful meetings which had just been held; crowds were being turned away from all the lectures and papers—a very encouraging sign indeed.

"He went on to say a few words about the financing of biochemistry. He was encouraged to do this because The Biochemical Society had already had a symposium devoted to this matter.

BIOCHEMISTRY FLOURISHING

"Biochemistry was certainly flourishing, but there were considerable gaps, and, in spite of anything that is said to the contary, we were not making the best use of our talent in this country. Talent was not so plentiful that we could afford to squander it in any branch of science. Biochemical departments were well established in many universities, but not in all, and of course there was now an opportunity to do better in some of the newly-established universities than in some of the older universities.

"Biochemistry should be all-pervasive, and there was room for professional biochemists in departments not labelled "Biochemistry". But there was one great difficulty—a practical difficulty—and that is that once a man got outside a biochemistry department there was no certainty, even if he were intelligent enough, of his getting a reasonably-paid job which would give him "tenure". That was to say a tenure such as professors have. This question of jobs with tenure seemed to be a very important matter.

"There were many discussions on the organization of science going on at the present time, and from all these discussions at least one thing emerged with crystal clarity, and that was that there were not enough senior jobs with adequate tenure to look after the young man who came in at the bottom and who, when he reached his 30s, was often left high and dry; and also, in spite of any words to the contrary, we were losing too many of our good people abroad from lack of facilities and lack of suitable jobs in this country.

"Sir Howard Florey said that he believed that scientific societies would, in the future, have to pay more attention to the organization of science, because they were the people who actually did the work. It would become increasingly important to think how best could results be obtained, because a vast amount of time was wasted raising money in order to do experiments. One of the most distinguished biologists of this country had said that during the last 10 years he had had to apply to 18 separate sources to keep his work going.

"The Biochemical Society had had a remarkable past and had amongst its members some of the world's most distinguished biochemists. Sir Howard ended by saying, "We all hope that you will have as renowned members in the future as you have at present and have had in the past. I, as one not a biochemist but as one who is working on what is commonly called the 'B' side and who had a profound admiration for your work—convey all the best wishes we can possibly give you for your future prosperity."

Sir Charles Harington

The Biochemical Society gave a Dinner in honour of Sir Charles Harington at the Hall of the Worshipful Society of Apothecaries of London on Friday, 5 October 1962. The toast of 'Our Guest of Honour' was proposed by the Chairman of the Committee, J. N. Davidson. The dinner marked the occasion of Sir Charles's retirement as Director of the National Institute for Medical Research. Reference was made to Harington's distinguished service as Editor of the *Journal* (see p. 63) and to his very important role in founding the International Union of Biochemistry. It was specially appropriate that Davidson who with Dickens (as Officers of the Society) helped to fight the battle for recognition of IUB should have had an opportunity to refer to the final successful outcome.

HONORARY MEMBERS

W. D. Halliburton was made an Honorary Member in November 1923 and when he died in 1931 he was the only Honorary Member. The next election was that of Sir Arthur Harden (November 1938). Sir Frederick Gowland Hopkins and R. H. A. Plimmer were elected in September 1943 and Sir Charles Martin was honoured in 1951. By 1958 the Society had no surviving honorary members. According to the rules then in force honorary members were deprived of the right to vote and the Committee proposed a new variation of the rule. It also recommended that Sir Henry Dale be invited to accept honorary membership. The rule was amended in 1958 to read as follows:

> "Honorary Members shall pay no subscription but shall receive the *Journal* and have all the privileges of membership of the Society including the right to vote".

In 1959 Sir Rudoph Peters was invited to accept honorary membership at the same time as he received the Hopkins Medal and in 1960 Sir Charles Harington was invited to accept honorary membership as a token of appreciation of the members of the Society for his services over so many years.

In 1964 a subcommittee was appointed to make recommendations about honorary membership. This committee reported:

1. The present practice of restricting Honorary Membership to a very limited number of specially distinguished biochemists is desirable. In view of the increasing membership of the Society, the number of Honorary Members might be increased a little, but the total should not exceed 10.

2. Scientific distinction should be the main requirement for election to Honorary Membership.

3. For the time being, Honorary Membership should be confined to members of the Society who are at or near retiring age.

4. A list of the names of Honorary Members shall be included in the Handbook of the Society. There shall also be published in the Handbook lists of the names of those who have received medals or lectureships from the Society during the period of 25 years preceding the date of publication of each edition of the Handbook.

In 1965 Sir John Gaddum was elected to honorary membership, Sir Charles Dodds, Professor R. A. Morton and Sir Robert Robinson were elected in 1966. Sir Hans Krebs and Professor F. Dickens became Honorary Members in 1967.

In 1969 it was decided that while Honorary Membership had been for some reason confined to Members residing in the U.K., consideration should be given in future to Members residing overseas, and consequently the maximum permitted number of Honorary Members was raised to 12. At the time of going to press, however, no overseas Member had been elected to Honorary Membership. It is hoped that this deficiency will soon be rectified.

MEDALS, LECTURES, FELLOWSHIPS AND CONFERENCES

In 1956 the Committee of the Society considered a request to establish a Memorial Lecture and a Medal to be presented every second year to a young British or French biochemist, the Medal being provided by the sponsors. After considerable discussion the Committee resolved "that since the institution of such a medal and lectureship would involve a completely new departure on the part of the Society, it could not agree to the request". But a seed had been sown. A somewhat different scheme soon gained committee support so that in 1958 the Society was able to institute the *Hopkins Memorial Lecture*, to be given every two years, generally in London to coincide with one of the major meetings of the Society. The Lecturer receives an honorarium and a bronze medal. He is expected "to assess the impact of recent advances in his chosen field upon progress in biochemistry". The lecture is followed by a dinner to which the surviving members of Sir Frederick's immediate family are invited. Those who have given the Lecture are listed on page 159.

In March 1961 a *Jubilee Lecture* was established to commemorate the Society's 50th anniversary. The Lecture is given each year in which there is no Hopkins Memorial Lecture, normally in London and also at a suitable centre outside London. The Lecturer receives an honorarium and is expected to lecture on his chosen field of research ((see p. 159)

In 1963 the Society was invited to administer a fund raised by a group of his associates in memory of David Keilin. The Committee regarded the invitation as an honour. A *Keilin Memorial Lecture* was duly instituted in January 1964. The Lecture is given every other year (see p. 159). As in the case of the *Hopkins Memorial Lecture*, the lecture is followed by a dinner, and the surviving members of Professor Keilin's immediate family are invited as the Society's guests.

The *Colworth Medal* was donated in 1963 by the Unilever Research Laboratory, Colworth House, Sharnbrook, Bedford, to be awarded annually to a British biochemist who shall not normally be over the age of 35. The recipient is expected to give a lecture at a meeting of the Society and at one of the Unilever Research Laboratories (see p. 160).

The *CIBA Medal and Prize* was donated in 1964 by CIBA Research Laboratories Ltd of Horsham, Sussex, for award each year in recognition of outstanding research in any branch of biochemistry. The award is for work carried out in Britain and Northern Ireland, but is open to candidates of any nationality. The Medal carries with it a prize of £50 and the recipient is expected to give a lecture.

BDH Chemicals Ltd, Poole have donated a medal and a prize of £200 to be awarded triennially by the Biochemical Society for outstanding work carried out in a laboratory situated in the British Isles or Eire and leading to advances in biochemistry related to the development and application of a new reagent or method. Nominees must be members of the Biochemical Society. The first award is to be made in 1969 and the recipient will be expected to give a lecture at the 500th Meeting of the Society in December 1969.

All these awards are made by *ad hoc* subcommittees of the Committee of the Society (or by a special committee) and the lectures are published in the *Biochemical Journal*. The Society has put on one side substantial funds to meet the costs of the *Hopkins Memorial Lecture* and the *Jubilee Lecture*.

FELLOWSHIPS

The Unilever organization has established two Fellowships of £1,500 per annum to be awarded each academic year, one for research in biochemistry in a laboratory in continental Europe by British citizens resident in Great Britain, and one for work in Britain by European nationals. These are post-doctoral awards tenable for one year in the first instance. It is a condition of the award of a Fellowship that it shall be acknowledged in any publication resulting from the work carried out during the period of the award, that the recipient was a 'Unilever European Fellow of the Biochemical Society' (see p. 000). Three Fellowships will be available for 1969–1970.

The Boehringer Corporation (London) Ltd has donated £500 to the Biochemical Society for a number of Travelling Fellowships in honour of Sir Hans Krebs. The object is to enable research students and other young biochemists to spend short periods in another laboratory or to attend a summer school or to acquire training and experience not available in their own country. Applicants must be members of the Biochemical Society and in principle should not be over the age of 30. The maximum grant under any one Fellowship will be £125.

H. R. V. Arnstein recalls that A. T. James, and through him H. Wilkinson, of Unilever Ltd, had a great deal to do with establishing the Colworth Medal and the Unilever European Fellowships. D. F. Elliott had a notable share in persuading the Director of the Ciba Laboratories to found the Ciba Medal. The initiative for the BDH Chemicals Prize came from the Company through C. R. Bayley. As Arnstein has remarked "these contributions are perhaps indicative of an increasing interest of industry in biochemistry and hence in the Society".

THE HARDEN CONFERENCES

In 1967 the sum of £4,000 became available to the Society under the Will of the late Sir Arthur Harden. The money was to be held in trust and the income used to defray the cost of publishing the results of original research in biochemistry. There was considerable discussion in Committee about the best means to use the money and legal advice was sought as to what was possible within the terms of the Will.

Among the suggestions made was the founding of annual small research conferences on the lines of the well known Gordon research Conferences in the U.S.A. The idea of conferences limited to less than 150 persons had considerable appeal to the Committee but the project had to be assessed apart from the legacy.

In September 1967 the Committee agreed to use the interest from the Harden legacy to support the *Biochemical Journal* and in February 1968 it was decided that instead of calling the new type conferences 'Biological Research Conferences' they should be named the 'Harden Conferences' to commemorate a man who had done so much for the *Journal* and the Society.

A special subcommittee was constituted and recommended that a conference under the title 'The Biological Role of Proteins' should be held at Wye College under the Chairmanship of Professor D. C. Phillips. The dates would be 14–19 September 1969 inclusive. The cost of the first Conference of this type was expected to exceed £3,000. The Society would set aside a sum equivalent to the income from the Harden legacy. The Chairman of the Committee had approached The Royal Society and The Wellcome Trust and both bodies had kindly agreed to make generous donations towards the Conferences. Further support was obtained from the Science Research Council, the Biological Council and Whatman Biochemicals Ltd.

SIR FREDERICK GOWLAND HOPKINS

British biochemists have on several occasions paid tribute to the inspiration of Sir Frederick Gowland Hopkins. A volume entitled Perspectives in Biochemistry was published by the Cambridge University Press in 1937 to celebrate Hopkins's seventy-fifth birthday. It consisted of thirty-one essays by 'past and present members of his laboratory' and was edited by J. Needham and D. E. Green. Apart from the value of 'a concerted creative effort' as a sign of gratitude and respect the volume has permanent merit as a picture of biochemistry in relation to general biology in the years just before the War.

Sir Frederick Gowland Hopkins, O.M., F.R.S. Acted in negotiations over publication of *Biochemical Journal*, First Chairman of Committee 1913–1914. Honorary Member 1943

Ernest Baldwin wrote a brief memoir *Gowland Hopkins* to celebrate the Centenary of Hopkins's birth in 1961. This was published by Van den Berghs Ltd as "a tribute from his friends in the margarine industry to one whose services to nutrition enriched the well-being of mankind". The fullest tribute came from J. Needham, delivering a Frederick Gowland Hopkins Centenary Lecture at Cambridge (20 November 1961) and later published through the generosity of the Burroughs Wellcome Fund in *Perspectives in Biology and Medicine* 1962 VI, 1.

On the occasion of the First International Congress of Biochemistry held in Cambridge in 1949 a memorial volume entitled *Hopkins and Biochemistry* edited by J. Needham and E. Baldwin was published and presented to members of the Congress by their Cambridge colleagues. The volume took the form of selected excerpts from Hopkins's writings with suitable commentaries.

At the request of the Council of the Royal Society the present writer organized a Symposium on Biochemistry and Nutrition held on 26 October 1961, "to celebrate the Centenary of the birth of a former President of the Royal Society, Sir Frederick Gowland Hopkins, O.M. (1861–1947)". The opening address was given by J. Needham. Other speakers in the first session went on to re-assess in perspective four topics which interested Hopkins, namely ascorbic acid (L. J. Harris), tryptophan (F. C. Happold) glutathione (N. W. Pirie) and pterins (E. Lester Smith). The remainder of the symposium was devoted to vitamins in nutrition (T. Moore); nutrition and growth (R. A. McCance and E. M. Widdowson); proteins in nutrition (B. S. Platt); protein metabolism in human protein malnutrition (J. C. Waterlow); vitamins A and B_{12} and some comments on refection (S. K. Kon); lipids and diet (R. Nicolaysen) and some aspects of lipid biochemistry (G. Popják).

Many memorable things were said in the Symposium but the present writer particularly recalls the contribution of T. Moore. Speaking about one of Hopkins's most famous papers (*J. Physiol.*, 1912, **44**, 425) Moore remarked:

"At this point, therefore, we must give Hopkins credit not only for making his discoveries, but also for convincing his contemporaries of their importance."

Moore however continued:

"Scientific honesty now requires mention of an aftermath of these famous experiments which undoubtedly involved Hopkins in considerable worry and perplexity . . . Other workers however could not confirm his findings. . . ."

Hopkins worked hard to answer such criticism and 33 years later (Hopkins and Leader, *J.Hyg.Camb.*, 1945, **44**, 149) its effects were still discernible. Moore concluded:

"in my own opinion we are still without an adequate and unequivocal explanation of why the experiments in 1912 gave such convincing results."

Admitting that this was a "knotty problem" Moore hazarded the explanation that Hopkins's basal diet was "not very highly purified by modern standards" and it may have produced "a partial rather than a complete deficiency of some vitamin'—responsive to small doses of milk".

"If he (Hopkins) were with us now perhaps we might ask him exactly what he meant in 1906 when he spoke of 'nutritive errors' which led to conditions less severe than scurvy or rickets but which affected the health of individuals to a degree most important to themselves."

Hopkins was not only a very distinguished pioneer he was also the creator of a great School of biochemistry, dedicated to the pursuit of the same aims as 'Hoppy' himself.

In 1921 the present writer was Secretary of the students' Chemical Society at Liverpool and Hopkins had agreed to give a lecture. A few days beforehand it was necessary to find a student willing to propose a vote of thanks but the ex-servicemen of the time knew little about 'Hoppy' or about biochemistry and the man 'drafted' agreed to act only if he was told what to say. Accordingly the Secretary took him to see W. Ramsden then Professor of Biochemistry. After he had understood the situation, Ramsden was most kind. He said that a first discovery might be largely accidental but that a scientist who made two discoveries was probably good while anybody who made three was certainly very good. He then explained *four* discoveries by Hopkins and made it clear that our visitor was 'a great man'. Time has not changed Ramsden's assessment.

F. A. Robinson writes: "Another event of importance that took place during my time as Treasurer was the suggestion that a medal should be provided for an Annual Lecture to commemorate Sir Frederick Gowland Hopkins. This was the first of the Society's commemoration medals. Towards the end of 1957 the sum of £2,000 was allocated for this purpose from the Society's fund and in April 1958 I visited Pinches, the medallists, who at that time occupied a minute building sandwiched between tall new office blocks on the South Bank of the River, and discussed the procedure for producing the Hopkins Medal. A week or two later I visited Cambridge and obtained from Mrs Holmes three photographs of her father from which Pinches prepared pencil sketches of "Hoppie", the final one being submitted to the family for approval. In November I inspected the plaster cast at Pinches, a thing the size of a dinner plate from which the medal was eventually prepared by a reduction process carried out 'on a gramophone'. The first Hopkins Lecturer was, of course, Sir Rudolph Peters, who received his medal from E. J. King at the meeting in April 1959."

DAVID KEILIN

David Keilin was a parasitologist and cell physiologist whose name will always be linked with the cytochrome system.

He was born in Moscow and educated at Warsaw, Liège and Paris. His interest in biology was aroused during his premedical studies at Liège. He was troubled by asthma and this led him to give up the medical course; he then studied philosophy in Paris under Bergson and biology under Caullery. Entomology proved a particular attraction and at the end of the course he took up research on the larval development of Diptera and discovered that the fly *Pollenia rudis* lays its eggs in the earthworm's sperm sacs. The developing larva later devours the worm, then buries itself in the soil, forming a puparium from which emerges the adult fly.

This and other remarkable discoveries gave Keilin a growing reputation as an entomologist.

Shortly after the outbreak of the 1914–1918 War Keilin was invited to become a research assistant to G. H. F. Nuttall at Cambridge. He accepted the post and later held a Beit Fellowship. Keilin worked at Cambridge from February 1915 until his death in 1963. He continued his entomological researches and discovered that the larva of the fly *Gasterophylus* which develops as a parasite in horses' stomachs contains haemoglobin. While searching in the adult fly for traces of haemoglobin with a microspectroscope he noticed in its wing muscle a four banded absorption spectrum, quite different from that of haemoglobin, which he then found to be present in all aerobic cells. The spectrum was due to a system of respiratory pigments which he designated cytochromes. Much of Keilin's subsequent research concerned biological oxidation with particular reference to the isolation, properties and roles of the different cytochromes and haematin proteins in general.

In 1921 Keilin moved with Professor Nuttall's group to the newly established Molteno Institute. In 1931 he succeeded Nuttall as Quick Professor of Biology and Director of the Institute. He continued to direct research in parasitology and through his various investigations especially those on cellular respiration and on metallo-enzymes he contributed greatly to the development of biological research in this country and abroad.

Keilin was awarded the Royal Society's Copley Medal in 1951. E. F. Hartree (*Biochem. J.*, 1963, **89**, 1) has written memorably about the early spectroscopic work and about Keilin's brilliant use of the microspectroscope. Keilin was generous in paying tribute to the earlier work of MacMunn (1884–1886). So far as possible he kept aloof from controversy himself and over the years he quietly and steadily developed his pioneer work on the cytochrome system. An account of Keilin's life and work written by T. Mann (Biographical Memoirs of Fellows of The Royal Society, Vol. 10, 1964) lays emphasis on contributions to the comparative biochemistry of haem pigments and to a range of enzyme problems. It also quotes a letter to *The Times* by M. F. Perutz:

"J. C. Kendrew and I owe Keilin a tremendous debt, for he was one of the first to see the potentialities of the physical approach to biochemistry. Until the 1950s we had no facilities for biochemical work in the Cavendish Laboratory; Keilin gave us a bench space in his institute even though he was short of space himself and he helped us to grow protein crystals. He also persuaded the Faculty Board to let me lecture on Molecular Biology long before that subject had even acquired its name. When Kendrew's research and mine were in danger of closing down for lack of support by the University Keilin suggested and supported Sir Lawrence Bragg's approach to the Medical Research Council which saved us.

But to me his most important gifts were his confidence in me and the warmth of his friendship which helped me to gain confidence in myself."

Keilin did not take much part in the affairs of the Biochemical Society but a substantial portion of his sustained collaborative work with E. F. Hartree appeared in the *Journal*. The Society was honoured by the invitation to sponsor the Keilin Memorial Lecture.

The cytochromes increase in number, complexity and significance as the years go by and many research schools are still striving to understand them more fully.

A valuable book entitled 'The History of Cell Respiration and Cytochrome' (Cambridge University Press 1966) by Professor Keilin was prepared for publication by his daughter Joan Keilin.

SIR ARTHUR HARDEN

Arthur Harden was born in Manchester on 12 October 1865, the son of a Manchester business man. He was brought up in an austere nonconformist atmosphere and was educated at Tettenhall College in Staffordshire. In January 1882 he entered the Owen's College, Manchester to study chemistry under Professor Roscoe, and in 1885 he graduated in the Victoria University with first class honours in chemistry. A year later he was awarded the Dalton scholarship.

Harden's first research was "The action of silicon tetrachloride on aromatic amide compounds" but then he proceeded to Erlangen and, under the direction of Otto Fischer, prepared β-nitroso-naphthylamine and investigated its properties. Here he was awarded the degree of Ph.D. after which he returned to Manchester, firstly as junior, and later as senior lecturer in chemistry under Professor H. B. Dixon. Harden remained at Manchester for another nine years, during which his activity seems to have been devoted chiefly to teaching and literary work.

In 1897 he was appointed as chemist to the Lister (then called the Jenner) Institute of Preventive Medicine in London. He had a wide knowledge of chemistry and had proved himself to be a successful teacher and became responsible for teaching the chemical course, which was mostly concerned with the analysis of water and foods, to medical practitioners desiring to become Medical Officers of Health. These courses were later superseded by special teaching for a diploma in public health conducted in London medical schools and Harden then found that he could devote himself fully to research. At the time Harden was in charge of the Chemical Department at the Institute, but in 1905 it was fused with the Biochemical Department and Harden was placed in charge of the composite department. In 1912, in recognition of his outstanding work on

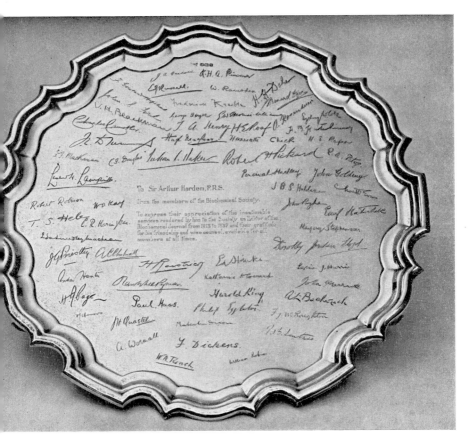

The silver salver presented by the Biochemical Society to Sir Arthur Harden, F.R.S. on the 11 March 1938 to mark the occasion of his retirement after 25 years from the Editorship of the *Biochemical Journal*

bacterial chemistry and alcoholic fermentation, he was made Professor of Biochemistry in the University of London.

It was during his earliest days at the Lister Institute that Harden started an investigation of the fermentation of sugars by bacteria. Subsequently he embarked on some ten years of research on alcoholic fermentation leading to the discovery of co-zymase and the essential role of phosphoric esters in such fermentation. Soon after these discoveries other workers found that phosphorylation provided the clue to many other biological phenomena, including the chemistry of muscle and bone.

During the First World War Harden was left in charge of the Lister Institute as Acting Director and since he wished to devote himself to a subject which would contribute to the war effort, he abandoned his researches on alcoholic fermentation and investigated instead two of the then known accessory food factors or vitamins, lack of which there was good reason to believe were responsible for the diseases beri-beri and scurvy respectively. Both diseases had occurred amongst troops in outposts in Africa and Asia.

Notable work was done at the Lister Institute under Harden's inspiration and encouragement and his wide knowledge was exercised in editing the three editions of the now classical reports on accessory food factors published by the Medical Research Council in 1919, 1924 and 1932.

Recognition of the importance of Harden's researches came from many quarters. In 1907 he was elected Fellow of the Royal Society, on the council of which he served from 1921 to 1923. In 1929 he shared the Nobel prize for chemistry with von Euler. The Universities of Manchester, Liverpool and Athens conferred honorary degrees upon him and the Kaiserlich Leopold Deutsche Akadamie der Naturforscher of Halle elected him to its membership.

Harden retired from the Lister Institute in 1930 and in the following year he was appointed to its governing body on which he served until his death in 1940. He became Emeritus Professor of Biochemistry at the University of London in 1931 and the Royal Society awarded him its Davy Medal in 1935. In 1936 he received the honour of knighthood.

Where the Biochemical Society, however, is concerned, Sir Arthur was noted for his length of service, 25 years, as one of the first two editors of the *Biochemical Journal*. He was appointed as such by the Biochemical Society immediately after the latter had acquired the *Journal* in 1913. During his appointment he had four co-editors in turn until he relinquished the post in 1937.

To mark the high respect and gratitude felt by all members of the Society, the Committee decided to offer him a distinctive token of remembrance. With Harden's concurrence it consisted of a silver salver bearing facsimile signatures of all those still living who had served with him on committees.

The presentation took place at the Annual General Meeting on 11 March 1938. The Chairman of Committee, H. J. Channon, in his address traced the growth curve of the number of words in the 25 volumes. Starting at 180,000 it had reached 1,500,000 per volume. The total in the period was some 18 million, all of which Harden read in proof. This is only one sign of the magnitude of his work. Sir Frederick Gowland Hopkins, who spoke afterwards, stressed the sacrifices Harden must have made in his time for his valuable researches on fermentation. Arthur Harden in his reply spoke of the function of editors. Expecting 'another enemy made' for any alteration he desired, he found instead 'there was another friend'. He expressed his gratitude to the Cambridge University Press for their care in the production of the *Journal*. The account of this occasion is in the *Journal* (1938) **32**, 769.

Sir Arthur was elected to the Honorary Membership of the Biochemical Society in 1938. Before his death he willed part of his estate to the Society: the income therefrom was to be applied in defraying the cost of publication of the results of original research in biochemistry.

ANNEXES

LIST OF TRUSTEES

1929 L. Baker
H. W. Dudley, succeeded in 1938 by A. C. Chibnall
J. A. Gardner, succeeded in 1946 by H. Raistrick
Sir Arthur Harden, succeeded in 1942 by J. C. Drummond
H. D. Kay
R. H. A. Plimmer
Sir Charles Harington
1938 A. C. Chibnall
1956 J. H. Bushill
Sir Rudolph Peters

HONORARY MEMBERS

1923 W. D. Halliburton
1938 Sir Arthur Harden
1943 Sir Frederick Gowland Hopkins
1943 R. H. A. Plimmer
1959 Sir Henry Dale
1959 Sir Rudolph Peters
1961 Sir Charles Harington
1965 Sir John Gaddum
1966 R. A. Morton
1966 Sir Charles Dodds
1966 Sir Robert Robinson
1967 Sir Hans Krebs
1967 F. Dickens

THE COMMITTEE

According to the rules of the Society the changes in the Committee have been carried out at each Annual General Meeting. At its first meeting after its formation in 1913 the Society decided to appoint a Chairman for each year. In 1921 the Chairman was asked to act as President on any special occasion. The Chairman was not necessarily the representative at other times. The Secretary or Treasurer or Editor was often the most suitable, or a member of the committee might be asked.

CHAIRMEN OF COMMITTEE (1911-1950)

1913–14	F. G. Hopkins	1932–33	C. G. L. Wolf
1914–15	W. M. Bayliss	1933–34	R. Robison
1915–16	V. H. Blackman	1934–35	F. L. Pyman
1916–17	G. Barger	1935–36	H. J. Page
1917–18	A. Harden	1936–37	P. Haas
1918–19	B. Dyer	1937–38	H. J. Channon
1919–20	W. M. Bayliss	1938–39	R. A. Peters
1920–21	P. Haas	1939–40	R. H. A. Plimmer
1921–22	S. B. Schryver	1940–41	G. M. Findlay
1922–23	R. H. A. Plimmer	1941–42	D. P. Cuthbertson
1923–24	J. C. Drummond	1942–43	J. C. Drummond
1924–25	P. Hartley	1943–44	J. Vargas Eyre
1925–26	H. W. Dudley	1944–45	E. C. Dodds
1926–27	C. Lovatt Evans	1945–46	A. C. Chibnall
1927–28	Ida Smedley Maclean	1946–47	F. A. Robinson
1928–29	R. A. Peters	1947–48	Margaret M. Murray
1929–30	T. S. Hele	1948–49	W. T. J. Morgan
1930–31	T. A. Henry	1949–50	H. Raistrick
1931–32	E. Hatschek		

HONORARY SECRETARIES (1911-1950)

1911–19	R. H. A. Plimmer
1919–22	J. C. Drummond
1922–24	H. W. Dudley
1924–27	P. Hartley
1927–29	H. D. Kay and R. Robison
1929–30	R. Robison and C. R. Harington
1930–38	A. C. Chibnall and H. Raistrick
1938–40	A. C. Chibnall and F. G. Young
1940–43	F. G. Young and W. T. J. Morgan
1943–45	W. T. J. Morgan and W. Robson
1945–46	W. Robson and D. P. Cuthbertson
1946–47	W. Robson and J. N. Davidson
1947–50	J. N. Davidson and J. H. Birkinshaw

HONORARY TREASURERS (1911-1952)

1913–1944   J. A. Gardner
1944–1952   J. H. Bushill

OFFICERS AND COMMITTEE 1950 ONWARDS

1949–1950

Chairman: H. Raistrick
Treasurer: J. H. Bushill
Secretaries: J. N. Davidson, J. H. Birkinshaw
Symposium Organiser: R. T. Williams

E. N. Allott	S. R. Elsden	M. Stacey
E. Baldwin	G. A. D. Haslewood	L. A. Stocken
F. Dickens	T. F. Macrae	Leslie Young
E. C. Dodds	W. T. J. Morgan	

1950–1951

Chairman: F. Dickens
Treasurer: J. H. Bushill
Secretaries: J. N. Davidson, L. Young
Symposium Organiser: R. T. Williams

E. N. Allott	S. R. Elsden	H. Raistrick
K. Bailey	G. A. D. Haslewood	L. A. Stocken
R. P. Cook	T. F. Macrae	W. V. Thorpe
E. C. Dodds	W. T. J. Morgan	

1951–1952

Chairman: E. C. Dodds
Treasurer: J. H. Bushill
Assistant Treasurer: F. A. Robinson
Secretaries: J. N. Davidson, L. Young
Symposium Organizer: R. T. Williams

E. N. Allott	S. R. Elsden	L. A. Stocken
K. Bailey	G. A. D. Haslewood	W. V. Thorpe
R. P. Cook	T. F. Macrae	F. G. Young
C. E. Dent	R. A. Peters	

1952–1953

Chairman: Sir Rudolph Peters
Treasurer: F. A. Robinson
Secretaries: L. Young, R. H. S. Thompson
Symposium Organizer: R. T. Williams

K. Bailey	S. J. Folley	L. A. Stocken
R. P. Cook	G. A. D. Haslewood	W. V. Thorpe
C. E. Dent	H. N. Munro	F. G. Young
S. R. Elsden	A. S. Parkes	

1953–1954

Chairman: F. G. Young
Treasurer: F. A. Robinson
Secretaries: R. H. S. Thompson, F. L. Warren
Symposium Organizer: R. T. Williams

K. Bailey	W. C. Evans	H. N. Munro
R. P. Cook	S. J. Folley	A. S. Parkes
E. M. Crook	T. W. Goodwin	Sir Rudolph Peters
C. E. Dent	E. Lester Smith	

1954–1955

Chairman: F. G. Young
Treasurer: F. A. Robinson
Secretaries: R. H. S. Thompson, F. L. Warren
Symposium Organizer: R. T. Williams

J. S. D. Bacon	T. W. Goodwin	H. N. Munro
E. M. Crook	J. K. Grant	A. S. Parkes
W. C. Evans	Sir Charles Harington	Sir Rudolph Peters
S. J. Folley	E. Lester Smith	

1955–1956

Chairman: Sir Charles Harington
Treasurer: F. A. Robinson
Secretaries: F. L. Warren, C. E. Dalgliesh
Symposium Organizer E. M. Crook

J. S. D. Bacon	S. J. Folley	E. Lester Smith
C. W. Carter	T. W. Goodwin	H. N. Munro
E. M. Crook	J. K. Grant	S. V. Perry
W. C. Evans	E. J. King	

1956–1957

Chairman: Sir Charles Harington
Treasurer: F. A. Robinson
Secretaries: F. L. Warren, C. E. Dalgliesh
Symposium Organizer E. M. Crook

J. S. D. Bacon	S. J. Folley	E. J. King
C. W. Carter	T. W. Goodwin	N. H. Martin
E. M. Crook	J. K. Grant	S. V. Perry
S. Dagley	T. S. G. Jones	G. R. Tristram
W. C. Evans		

1957–1958

Chairman: E. J. King
Treasurer: F. A. Robinson
Secretaries: F. L. Warren, C. E. Dalgliesh
Symposium Organizer: E. M. Crook

J. S. D. Bacon	J. K. Grant	S. V. Perry
P. N. Campbell	T. S. G. Jones	R. R. Porter
C. W. Carter	R. A. Morton	G. R. Tristram
S. Dagley	S. M. Partridge	

1958–1959

Chairman: E. J. King
Treasurer: F. A. Robinson
Secretaries: C. E. Dalgliesh, P. N. Campbell
Symposium Organizer: J. K. Grant

C. W. Carter	T. S. G. Jones	S. M. Partridge
S. Dagley	W. O. Kermack	S. V. Perry
K. S. Dodgson	H. McIlwain	R. R. Porter
G. N. Jenkins	R. A. Morton	G. R. Tristram

1959–1960

Chairman: R. A. Morton
Treasurer: F. A. Robinson
Secretaries: C. E. Dalgliesh, W. J. Whelan
P. N. Campbell
Symposium Organizer: J. K. Grant

S. Dagley	T. S. G. Jones	R. R. Porter
J. N. Davidson	W. O. Kermack	W. E. Van Heyningen
K. S. Dodgson	H. McIlwain	C. P. Whittaker
G. N. Jenkins	S. M. Partridge	

1960–61

Chairman: R. A. Morton
Treasurer: F. A. Robinson
Secretaries: P. N. Campbell, W. J. Whelan
Symposium Organizer: J. K. Grant

E. P. Abraham	G. N. Jenkins	R. R. Porter
S. P. Datta	N. F. MacLagan	H. J. Rogers
J. N. Davidson	R. Markham	W. V. Thorpe
K. S. Dodgson	H. McIlwain	W. E. Van Heyningen
Q. H. Gibson	S. M. Partridge	V. P. Whittaker
W. O. Kermack		

1961–1962

Chairman: J. N. Davidson
Treasurer: F. A. Robinson
Secretaries: P. N. Campbell, W. J. Whelan
Symposium Organizer: J. K. Grant

J. A. V. Butler	F. C. Happold	H. McIlwain
S. P. Datta	W. O. Kermack	W. E. Van Heyningen
K. S. Dodgson	E. Kodicek	V. P. Whittaker
Q. H. Gibson	N. F. MacLagan	

1962–1963

Chairman: J. N. Davidson
Treasurer: W. F. J. Cuthbertson
Secretaries: H. R. V. Arnstein, P. N. Campbell
Symposium Organizer: J. K. Grant

J. A. V. Butler	F. C. Happold	G. A. Snow
S. P. Datta	E. Kodicek	W. E. Van Heyningen
Q. H. Gibson	N. F. MacLagan	V. P. Whittaker
T. W. Goodwin	Helen K. Porter	

1963–1964

Chairman: F. C. Happold
Treasurer: W. F. J. Cuthbertson
Secretaries: H. R. V. Arnstein, P. N. Campbell
Symposium Organizer: J. K. Grant

G. S. Boyd	P. W. Kent	G. A. Snow
S. P. Datta	E. Kodicek	W. E. Van Heyningen
J. Glover	N. F. MacLagan	S. G. Waley
T. W. Goodwin	Helen K. Porter	V. P. Whittaker

1964–1965

Chairman: F. C. Happold
Treasurer: W. F. J. Cuthbertson
Committee Secretary: H. R. V. Arnstein
Meetings Secretary: K. S. Dodgson
International Secretary: W. J. Whelan
Symposium Organizer: T. W. Goodwin

W. N. Aldridge	P. T. Grant	H. J. Rogers
J. Baddiley	P. W. Kent	G. A. Snow
D. N. Baron	E. Kodicek	S. G. Waley
G. S. Boyd	Rosalind V. Pitt-Rivers	V. P. Whittaker
J. Glover	Helen K. Porter	

1965–1966

Chairman: Helen K. Porter
Treasurer: W. F. J. Cuthbertson
Committee Secretary: H. R. V. Arnstein
Meetings Secretary: K. S. Dodgson
International Secretary: W. J. Whelan
Symposium Organizer: T. W. Goodwin

W. N. Aldridge	P. T. Grant	A. Neuberger
J. Baddiley	B. S. Hartley	Rosalind V. Pitt-Rivers
D. N. Baron	P. W. Kent	G. A. Snow
G. S. Boyd	H. L. Kornberg	R. Whittam
J. Glover	J. Mandelstam	

1966–1967

Chairman: Helen K. Porter
Treasurer: W. F. J. Cuthbertson
Committee Secretary: H. R. V. Arnstein
Meetings Secretary: K. S. Dodgson
International Secretary: W. J. Whelan
Symposium Organizer: T. W. Goodwin

W. N. Aldridge	D. F. Evered	A. T. James
J. Baddiley	J. Glover	P. W. Kent
D. N. Baron	P. T. Grant	H. L. Kornberg
G. S. Boyd	H. Gutfreund	A. Neuberger
F. Dickens	B. S. Hartley	B. Spencer
S. R. Elsden	E. R. Hartree	

1967–1968

Chairman: A. Neuberger
Treasurer: W. F. J. Cuthbertson
Committee Secretary: K. S. Dodgson
Meetings Secretary: A. N. Davison
International Secretary: A. P. Mathias
Symposium Organizer: T. W. Goodwin

W. N. Aldridge	S. R. Elsden	A. T. James
H. R. V. Arnstein	D. F. Evered	H. L. Kornberg
J. Baddiley	P. T. Grant	H. A. Krebs
G. R. Barker	H. Gutfreund	R. M. S. Smellie
D. N. Baron	B. S. Hartley	R. H. S. Thompson
R. V. Brooks	G. A. D. Haslewood	

1968–1969

Chairman: A. Neuberger
Treasurer: W. F. J. Cuthbertson
Committee Secretary: K. S. Dodgson
Meetings Secretary: A. N. Davison
International Secretary: A. P. Mathias
Symposium Organizer: T. W. Goodwin

W. N. Aldridge	S. R. Elsden	I. Helen M. Muir
H. R. V. Arnstein	D. F. Evered	C. Ó hEocha
J. S. D. Bacon	G. A. D. Haslewood	D. C. Phillips
G. R. Barker	A. T. James	J. R. Quayle
D. N. Baron	H. L. Kornberg	R. M. S. Smellie
W. Bartley	H. A. Krebs	R. H. S. Thompson

EDITORS OF THE BIOCHEMICAL JOURNAL

Two editors and a representative editorial committee were appointed to edit the *Biochemical Journal* in 1913. The editorial committee was:

E. F. Armstrong F. Keeble
V. H. Blackman B. Moore
A. J. Brown W. Ramsden
J. A. Gardner E. J. Russell
F. G. Hopkins

G. Baxter replaced Armstrong in 1918 on his resignation. No successors were appointed to A. J. Brown and B. Moore. These names appeared on the title pages of the *Journal*, Vols. 7–32.

The success of the Society and the *Biochemical Journal* must be chiefly attributed to the first editors, especially A. Harden, who had as his first colleague, W. M. Bayliss, then H. W. Dudley, and afterwards C. R. Harington. F. J. W. Roughton assisted for three years with the papers on statistical questions and physical chemistry. Their periods of office were:

W. M. Bayliss and A. Harden, 1913–24
A. Harden and H. W. Dudley, 1924–30
A. Harden and C. R. Harington, 1930–34
A. Harden and C. R. Harington and F. J. W. Roughton, 1935–37

The *Journal* was edited from 1938 to 1942 by C. R. Harington with three associate editors, S. J. Cowell, F. Dickens, F. J. W. Roughton.

An Editorial Board was constituted in 1943 with one member as Chairman or chief to co-ordinate the work and be the responsible officer. In this way the heavy burden for the chief editor of reading and editing the numerous papers could be lightened. To date the Editorial Board have been:

1942–1943

F. G. Young (*Chairman*)
S. J. Cowell R. A. Kekwick B. C. J. G. Knight
F. Dickens E. J. King A. R. Todd

1943–1944

F. G. Young (*Chairman*)
F. Dickens B. C. J. G. Knight A. R. Todd
E. J. King J. R. P. O'Brien

1944–1945

F. G. Young (*Chairman*)

K. Bailey	E. J. King	J. R. P. O'Brien
F. Dickens	B. C. J. G. Knight	A. R. Todd

1945–1946

F. G. Young (*Chairman*)

K. Bailey	B. C. J. G. Knight	C. Rimington
F. Dickens	S. K. Kon	A. R. Todd
E. J. King	J. R. P. O'Brien	

1946–1947

E. J. King (*Chairman*)

K. Bailey	S. K. Kon	A. Neuberger
R. K. Callow	H. McIlwain	J. R. P. O'Brien
G. D. Greville	R. A. Morton	C. Rimington
B. C. J. G. Knight		

1947–1948

E. J. King (*Chairman*)

K. Bailey	H. McIlwain	J. R. P. O'Brien
R. K. Callow	R. A. Morton	C. Rimington
G. D. Greville	A. Neuberger	A. Wormall
S. K. Kon		

1948–1949

E. J. King (*Chairman*)

R. K. Callow	H. McIlwain	J. R. P. O'Brien
G. D. Greville	R. A. Morton	R. H. S. Thompson
R. A. Kekwick	A. Neuberger	A. Wormall
M. G. Macfarlane		

1949–1950

E. J. King (*Chairman*)

R. K. Callow	Marjorie G. MacFarlane	A. Neuberger
G. D. Greville	H. McIlwain	R. H. S Thompson
R. A. Kekwick	R. A. Morton	A. Wormall

1950–1951

E. J. King (*Chairman*)

R. K. Callow	Marjorie G. MacFarlane	R. H. S. Thompson
G. D. Greville	R. A. Morton	T. S. Work
R. A. Kekwick	A. Neuberger	A. Wormall

1951–1952
E. J. King (*Chairman*)

R. K. Callow	Marjorie G. MacFarlane	R. L. M. Synge
D. Herbert	R. A. Morton	T. S. Work
R. A. Kekwick	A. Neuberger	A. Wormall
W. Klyne	A. Pirie	

1952–1953
A. Neuberger (*Chairman*)

R. K. Callow	C. Long	A. G. Ogston
R. E. Davies	Marjorie G. MacFarlane	A. Pirie
D. Herbert	N. H. Martin	R. L. M. Synge
R. A. Kekwick	R. A. Morton	T. S. Work
W. Klyne		

1953–1954
A. Neuberger (*Chairman*)

R. K. Callow	C. Long	A. Pirie
R. E. Davies	Marjorie G. MacFarlane	R. L. M. Synge
S. R. Elsden	T. Mann	W. V. Thorpe
D. Herbert	N. H. Martin	T. S. Work
W. Klyne	A. G. Ogston	

1954–1955
A. Neuberger (*Chairman*)

H. R. V. Arnstein	C. Long	A. G. Ogston
R. E. Davies	Marjorie G. MacFarlane	R. L. M. Synge
S. R. Elsden	T. Mann	W. V. Thorpe
Q. H. Gibson	R. Markham	T. S. Work
W. Klyne	N. H. Martin	W. J. Whelan

1955–1956
A. G. Ogston (*Chairman*)   T. Work (*Deputy Chairman*)

E. P. Abraham	D. C. Harrison	D. M. Needham
H. R. V. Arnstein	C. Long	H. J. Rogers
S. R. Elsden	T. Mann	W. V. Thorpe
Q. H. Gibson	R. Markham	W. J. Whelan
C. H. Gray	N. H. Martin	

1956–1957

A. G. Ogston (*Chairman*)   T. Work (*Deputy Chairman*)

E. P. Abraham	D. C. Harrison	D. M. Needham
H. R. V. Arnstein	C. Long	H. J. Rogers
S. R. Elsden	T. Mann	W. V. Thorpe
Q. H. Gibson	R. Markham	W. J. Whelan
C. H. Gray	N. H. Martin	

1957–1958

A. G. Ogston (*Chairman*)   H. J. Rogers (*Deputy Chairman*)

E. P. Abraham	C. H. Gray	R. J. Pennington
H. R. V. Arnstein	D. C. Harrison	J. J. Scott
W. Bartley	C. Long	I. D. E. Storey
S. R. Elsden	L. W. Mapson	W. V. Thorpe
Q. H. Gibson	R. Markham	W. J. Whelan

1958–1959

A. G. Ogston (*Chairman*)   H. J. Rogers (*Deputy Chairman*)

E. P. Abraham	Q. H. Gibson	R. J. Pennington
W. Bartley	C. H. Gray	J. J. Scott
E. A. Dawes	D. C. Harrison	W. V. Thorpe
R. M. C. Dawson	L. W. Mapson	W. J. Whelan
G. E. Francis	R. Markham	

1959–1960

W. V. Thorpe (*Chairman*)   H. J. Rogers (*Deputy Chairman*)

E. P. Abraham	G. E. Francis	L. W. Mapson
W. N. Aldridge	Q. H. Gibson	R. Markham
W. Bartley	C. H. Gray	R. J. Pennington
E. A. Dawes	D. C. Harrison	J. J. Scott
R. M. C. Dawson	June Lascelles	

1960–1961

W. V. Thorpe (*Chairman*)   H. J. Rogers (*Deputy Chairman*)

E. P. Abraham	E. A. Dawes	June Lascelles
W. N. Aldridge	R. M. C. Dawson	J. Mandelstam
J. S. D. Bacon	C. H. Gray	L. W. Mapson
W. Bartley	D. C. Harrison	R. Markham
K. Burton	E. F. Hartree	J. J. Scott

1961–1962

W. V. Thorpe (*Chairman*)   H. J. Rogers (*Deputy Chairman*)

W. N. Aldridge	L. Fowden	J. Mandelstam
J. S. D. Bacon	H. Gutfreund	W. S. Peart
W. Bartley	D. C. Harrison	R. R. Porter
K. Burton	E. F. Hartree	P. J. Randle
E. A. Dawes	June Lascelles	S. G. Waley
R. M. C. Dawson		

1962–1963

W. V. Thorpe (*Chairman*)   H. J. Rogers (*Deputy Chairman*)

W. N. Aldridge	E. A. Dawes	J. Mandelstam
J. S. D. Bacon	L. Fowden	W. S. Peart
W. Bartley	H. Gutfreund	R. R. Porter
R. V. Brooks	E. F. Hartree	P. J. Randle
K. Burton	June Lascelles	S. G. Waley

1963–1964

H. J. Rogers (*Chairman*)   S. G. Waley (*Deputy Chairman*)

W. N. Aldridge	E. A. Dawes	J. Mandelstam
J. S. D. Bacon	L. Fowden	W. S. Peart
W. Bartley	H. Gutfreund	R. R. Porter
R. V. Brooks	E. F. Hartree	P. J. Randle
K. Burton	June Lascelles	D. N. Rhodes

1964–1965

H. J. Rogers (*Chairman*)   S. G. Waley (*Deputy Chairman*)

W. N. Aldridge	June Lascelles	R. R. Porter
R. V. Brooks	J. Mandelstam	D. N. Rhodes
J. B. Chappell	D. J. Manners	D. B. Roodyn
E. A. Dawes	I. Helen M. Muir	D. G. Walker
H. Gutfreund	W. S. Peart	R. Whittam
E. F. Hartree		

1965–1966

W. N. Aldridge (*Chairman*)   R. Whittam (*Deputy Chairman*)

R. V. Brooks	June Lascelles	R. R. Porter
J. B. Chappell	J. Mandelstam	D. N. Rhodes
H. Gutfreund	D. J. Manners	D. B. Roodyn
E. F. Hartree	I. Helen M. Muir	D. G. Walker

1966–1967
W. N. Aldridge (*Chairman*)
H. Gutfreund, J. R. Quayle, V. P. Whittaker (*Deputy Chairmen*)

J. L. Bailey	P. N. Magee	E. R. Redfern
R. V. Brooks	D. J. Manners	D. N. Rhodes
R. H. Burdon	J. H. Moore	D. S. Robinson
R. A. Cox	C. J. O. R. Morris	D. B. Roodyn
M. J. Crumpton	I. Helen M. Muir	R. E. Strange
D. F. Elliott	C. A. Pasternak	J. R. Tata
J. R. S. Fincham	A. R. Peacocke	P. K. Tubbs
J. N. Hawthorne	H. R. Perkins	D. G. Walker
J. R. Knowles	R. R. Porter	

1967–1968
W. N. Aldridge (*Chairman*)
J. R. Quayle, D. G. Walker, V. P. Whittaker (*Deputy Chairmen*)

M. L. Birnstiel	P. N. Magee	D. N. Rhodes
R. V. Brooks	D. J. Manners	D. S. Robinson
R. H. Burdon	J. H. Moore	R. Rodnight
R. A. Cox	C. J. O. R. Morris	D. B. Roodyn
M. J. Crumpton	I. Helen M. Muir	R. E. Strange
D. F. Elliott	A. C. T. North	J. R. Tata
P. B. Garland	C. A. Pasternak	P. K. Tubbs
J. N. Hawthorne	A. R. Peacocke	D. C. Watts
J. R. Knowles	H. R. Perkins	F. R. Whatley

1968–1969
W. N. Aldridge (*Chairman*)
D. S. Robinson, J. R. Quayle, D. G. Walker (*Deputy Chairmen*)

M. L. Birnstiel	J. D. Judah	H. R. Perkins
R. V. Brooks	J. R. Knowles	G. K. Radda
R. H. Burdon	P. N. Magee	D. N. Rhodes
R. A. Cox	D. J. Manners	R. Rodnight
Jill E. Cremer	J. H. Moore	D. B. Roodyn
M. J. Crumpton	C. J. O. R. Morris	R. E. Strange
D. F. Elliott	I. Helen M. Muir	J. R. Tata
P. B. Garland	A. C. T. North	P. K. Tubbs
B. S. Hartley	C. A. Pasternak	D. C. Watts
J. N. Hawthorne	A. R. Peacocke	F. R. Whatley

ADVISORY COMMITTEE FOR PUBLICATIONS

(The Advisory Committee for Publications was set up in February 1963 to "consider and make recommendations to the Committee of the Society on all matters concerning the publications of the Society with the exception of *Clinical Science*".)

V. P. Whittaker (*Chairman*)   P. N. Campbell (*Secretary*)
K. Burton   F. C. Happold   R. R. Porter
W. F. J. Cuthbertson   J. Mandelstam   H. J. Rogers

1964–1965
V. P. Whittaker (*Chairman*)   H. R. V. Arnstein (*Secretary*)
P. N. Campbell   E. F. Hartree   R. R. Porter
W. F. J. Cuthbertson   J. Mandelstam   H. J. Rogers
F. C. Happold

1965–1967
T. W. Goodwin (*Chairman*)   H. R. V. Arnstein (*Secretary*)
W. N. Aldridge   H. Gutfreund   R. R. Porter
P. N. Campbell   P. W. Kent   D. G. Walker
W. F. J. Cuthbertson   Helen K. Porter

1967–1969
T. W. Goodwin (*Chairman*)   K. S. Dodgson (*Secretary*)
W. N. Aldridge   C. J. O. R. Morris   P. K. Tubbs
P. N. Campbell   A. Neuberger   D. G. Walker
W. F. J. Cuthbertson   R. M. S. Smellie

SIR FREDERICK GOWLAND HOPKINS MEMORIAL LECTURERS

1958	Sir Rudolph Peters	1965	A. Szent-Gyorgi
1960	A. Neuberger	1967	H. A. Barker
1961	Sir Hans Krebs	1969	F. J. W. Roughton
1963	L. F. Leloir		

JUBILEE LECTURERS

1962	P. C. Zamecnik	1966	F. Lynen
1964	E. Lederer	1968	H. G. Khorana

DAVID KEILIN MEMORIAL LECTURERS

1964	A. Lwoff	1969	M. Eigen
1966	B. Chance		

THE COLWORTH MEDALLISTS

1963 H. L. Kornberg
1964 J. R. Tata
1965 J. B. Chappell

1966 M. R. Richmond
1967 L. J. Morris
1968 P. B. Garland

THE CIBA MEDALLISTS

1965 J. W. Cornforth ⎱ Joint award
     G. J. Popják ⎰
1966 R. R. Porter

1967 D. M. Blow
1968 W. J. Whelan

UNILEVER EUROPEAN FELLOWSHIPS

1965–1966 F. Novello Paglianti (Padua)
1966 J. H. Ottaway (Edinburgh)
1966–1967 M. I. Gurr (Birmingham and Harvard)
1967–1968 J. Barber (East Anglia)
          L. Mircevová (Prague)
          J. Stahl (Berlin)
1968–1969 E. Reiner (Belgrade)

BDH CHEMICALS LTD AWARD IN ANALYTICAL BIOCHEMISTRY

1969 B. S. Hartley

INTERNATIONAL MEETINGS

INTERNATIONAL CONGRESS OF BIOCHEMISTRY

Cambridge 1949
Paris 1952
London 1955
Brussels 1958

Moscow 1961
New York 1964
Tokyo 1967

FEDERATION OF EUROPEAN BIOCHEMICAL SOCIETIES

London March 1964
Vienna April 1965
Warsaw April 1966

Oslo July 1967
Prague July 1968
Madrid April 1969